CONGRESSIONAL POLICIES, PRACTICES AND PROCEDURES

ACTIONS NEEDED FOR SUCCESSFUL ENUMERATION OF THE 2020 CENSUS

CONGRESSIONAL POLICIES, PRACTICES AND PROCEDURES

Additional books and e-books in this series can be found on Nova's website under the Series tab.

CONGRESSIONAL POLICIES, PRACTICES AND PROCEDURES

ACTIONS NEEDED FOR SUCCESSFUL ENUMERATION OF THE 2020 CENSUS

BRYANT SCHNEIDER
EDITOR

Copyright © 2019 by Nova Science Publishers, Inc.

All rights reserved. No part of this book may be reproduced, stored in a retrieval system or transmitted in any form or by any means: electronic, electrostatic, magnetic, tape, mechanical photocopying, recording or otherwise without the written permission of the Publisher.

We have partnered with Copyright Clearance Center to make it easy for you to obtain permissions to reuse content from this publication. Simply navigate to this publication's page on Nova's website and locate the "Get Permission" button below the title description. This button is linked directly to the title's permission page on copyright.com. Alternatively, you can visit copyright.com and search by title, ISBN, or ISSN.

For further questions about using the service on copyright.com, please contact:
Copyright Clearance Center
Phone: +1-(978) 750-8400 Fax: +1-(978) 750-4470 E-mail: info@copyright.com

NOTICE TO THE READER

The Publisher has taken reasonable care in the preparation of this book, but makes no expressed or implied warranty of any kind and assumes no responsibility for any errors or omissions. No liability is assumed for incidental or consequential damages in connection with or arising out of information contained in this book. The Publisher shall not be liable for any special, consequential, or exemplary damages resulting, in whole or in part, from the readers' use of, or reliance upon, this material. Any parts of this book based on government reports are so indicated and copyright is claimed for those parts to the extent applicable to compilations of such works.

Independent verification should be sought for any data, advice or recommendations contained in this book. In addition, no responsibility is assumed by the Publisher for any injury and/or damage to persons or property arising from any methods, products, instructions, ideas or otherwise contained in this publication.

This publication is designed to provide accurate and authoritative information with regard to the subject matter covered herein. It is sold with the clear understanding that the Publisher is not engaged in rendering legal or any other professional services. If legal or any other expert assistance is required, the services of a competent person should be sought. FROM A DECLARATION OF PARTICIPANTS JOINTLY ADOPTED BY A COMMITTEE OF THE AMERICAN BAR ASSOCIATION AND A COMMITTEE OF PUBLISHERS.

Additional color graphics may be available in the e-book version of this book.

Library of Congress Cataloging-in-Publication Data

ISBN: 978-1-53616-716-0

Published by Nova Science Publishers, Inc. † New York

CONTENTS

Preface		vii
Chapter 1	2020 Census: Actions Needed to Address Challenges to Enumerating Hard-to-Count Groups *United States Government Accountability Office*	1
Chapter 2	2020 Census: Actions Needed to Address Key Risks to a Successful Enumeration *Robert Goldenkoff and Nick Marinos*	35
Chapter 3	2020 Census: Progress Report on the Census Bureau's Efforts to Contain Enumeration Costs *Robert Goldenkoff and Carol R. Cha*	83
Chapter 4	2020 Census: Actions Needed to Mitigate Key Risks Jeopardizing a Cost-Effective and Secure Enumeration *United States Government Accountability Office*	103
Index		145
Related Nova Publications		151

PREFACE

The Bureau is responsible for conducting a complete and accurate decennial census of the U.S. population. The decennial census is mandated by the Constitution and provides vital data for the nation. A goal for the 2020 Census is to count everyone once, only once, and in the right place. Achieving a complete and accurate census is becoming an increasingly complex task, in part because the nation's population is growing larger, more diverse, and more reluctant to participate. When the census misses a person who should have been included, it results in an undercount. Historically, certain sociodemographic groups have been undercounted in the census, which is particularly problematic given the many uses of census data. Chapter 1 reviews the Bureau's plans for enumerating hard-to-count groups in the 2020 Census. The remaining chapters report on the reasons the 2020 Census remains on the High-Risk List and the steps the Bureau needs to take to mitigate risks to a successful census.

Chapter 1 - A goal for the 2020 Census is to count everyone once, only once, and in the right place. Achieving a complete and accurate census is becoming an increasingly complex task, in part because the nation's population is growing larger, more diverse, and more reluctant to participate. When the census misses a person who should have been included, it results in an undercount. Historically, certain sociodemographic groups have been undercounted in the census, which is particularly problematic given the many uses of census data. GAO was asked to review the Bureau's plans for enumerating hard-to-count groups in the 2020 Census. This chapter examines (1) the Bureau's plans for improving the enumeration of the hard-to-count in 2020, and how that compares with 2010; and (2) the challenges the Bureau faces in improving the enumeration of the hard-to-count in 2020. GAO reviewed Bureau planning, budget, operational, and evaluation documents as well as documents of the hard-to-count related working groups of the Bureau's National Advisory Committee; and interviewed Bureau officials.

Chapter 2 - The Bureau is responsible for conducting a complete and accurate decennial census of the U.S. population. The decennial census is mandated by the Constitution and provides vital data for the nation. A complete count of the nation's population is an enormous undertaking as the Bureau seeks to control the cost of the census, implement operational innovations, and use new and modified IT systems. In recent years, GAO has identified challenges that raise serious concerns about the Bureau's ability to conduct a cost-effective count. For these reasons, GAO added the 2020 Census to its High-Risk list in February 2017. GAO was asked to testify about the reasons the 2020 Census remains on the High-Risk List and the steps the Bureau needs to take to mitigate risks to a successful census. To do so, GAO summarized its prior work regarding the Bureau's planning efforts for the 2020 Census. GAO also included preliminary observations from its ongoing work examining the IT systems readiness and cybersecurity for the 2020 Census. This information is related to, among other things, the Bureau's progress in developing and testing key systems and the status of cybersecurity risks.

Chapter 3 - At $13 billion, 2010's headcount was the costliest in U.S. history. Thus, over the next few years, the fundamental challenge facing Bureau leadership will be designing and implementing a census that controls the cost of the enumeration while maintaining its accuracy. This chapter focuses on progress the Bureau is making in three areas key to a more cost-effective enumeration: (1) transforming the Bureau into a higher-performing organization; (2) improving the cost-effectiveness of census-taking operations; and (3) strengthening IT management and security practices. This chapter is based on completed work that included an analysis of Bureau documents, interviews with Bureau officials, and field observations of census operations in urban and rural locations across the country.

Chapter 4 - One of the Bureau's most important functions is to conduct a complete and accurate decennial census of the U.S. population. The decennial census is mandated by the Constitution and provides vital data for the nation. A complete count of the nation's population is an enormous undertaking as the Bureau seeks to control the cost of the census, implement operational innovations, and use new and modified IT systems. In recent years, GAO has identified challenges that raise serious concerns about the Bureau's ability to conduct a cost-effective count. For these reasons, GAO added the 2020 Census to its high-risk list in February 2017. GAO was asked to testify about the Bureau's progress in preparing for the 2020 Census. To do so, GAO summarized its prior work regarding the Bureau's planning efforts for the 2020 Census. GAO also included preliminary observations from its ongoing work examining the 2018 End-to-End Test. This information is related to, among other things, progress on key systems to be used for the 2018 End-to-End Test, including the status of IT security assessments, and efforts to update the life-cycle cost estimate.

In: Actions Needed ...
Editor: Bryant Schneider

ISBN: 978-1-53616-716-0
© 2019 Nova Science Publishers, Inc.

Chapter 1

2020 CENSUS: ACTIONS NEEDED TO ADDRESS CHALLENGES TO ENUMERATING HARD-TO-COUNT GROUPS[*]

United States Government Accountability Office

ABBREVIATIONS

Bureau	U.S. Census Bureau
Recovery Act	The American Recovery and Reinvestment Act of 2009

---

[*] This is an edited, reformatted and augmented version of the United States Government Accountability Office Report to Congressional Requesters, Publication No. GAO-18-599, dated July 2018.

WHY GAO DID THIS STUDY

A goal for the 2020 Census is to count everyone once, only once, and in the right place. Achieving a complete and accurate census is becoming an increasingly complex task, in part because the nation's population is growing larger, more diverse, and more reluctant to participate. When the census misses a person who should have been included, it results in an undercount. Historically, certain sociodemographic groups have been undercounted in the census, which is particularly problematic given the many uses of census data.

GAO was asked to review the Bureau's plans for enumerating hard-to-count groups in the 2020 Census. This chapter examines (1) the Bureau's plans for improving the enumeration of the hard-to-count in 2020, and how that compares with 2010; and (2) the challenges the Bureau faces in improving the enumeration of the hard-to-count in 2020. GAO reviewed Bureau planning, budget, operational, and evaluation documents as well as documents of the hard-to-count related working groups of the Bureau's National Advisory Committee; and interviewed Bureau officials.

WHAT GAO RECOMMENDS

GAO recommends that the Bureau take steps to ensure that forthcoming changes and decisions on its hard-to-count related efforts are integrated with other operational efforts and that it collects data on its 2020 partnership hiring efforts.

The Department of Commerce agreed with GAO's recommendations, and the Bureau provided technical comments that were incorporated, as appropriate.

WHAT GAO FOUND

The Census Bureau's (Bureau) plans for enumerating groups considered hard-to-count, such as minorities, renters, and young children, in the 2020 Census includes the use of both traditional and enhanced initiatives. For example, the Bureau plans to continue using certain outreach efforts used in 2010, such as a communications campaign with paid advertising, partnerships with local organizations, and targeted outreach to immigrant and faith-based organizations. The Bureau also plans enhancements to its outreach efforts compared to 2010. For example, to help address the undercount of young children, the Bureau revised the census questionnaire and instructions to enumerators to more explicitly include grandchildren in counts. Other planned changes include:

- Expanded languages: The Bureau plans to offer more non-English language response options and instructional materials than for 2010.
- More partnership specialists: The Bureau plans to hire nearly twice as many partnership specialists as it had planned for the 2010 Census to recruit partner organizations in local communities.
- Earlier partnership hiring: The Bureau started hiring a small number of partnership staff in October 2015—2 years earlier than it did for 2010.

While efforts have been made, enumerating hard-to-count persons in 2020 will not be easy. Aside from the inherent difficulties of counting such individuals, the Bureau faces certain management challenges related to its hard-to-count efforts.

- First, the Bureau's hard-to-count efforts are distributed across over one third of its 35 operations supporting the 2020 Census. And while decentralized operations can provide flexibility, to enhance visibility over these hard-to-count efforts, the Bureau recently developed a draft operational document. However, the Bureau will continue to face challenges in ensuring its hard-to-count efforts integrate with each other. For example, some of the detailed plans for 10 of the hard-to-count efforts were released in 2016 and are awaiting updates, while 4 plans have yet to be released. With less than 2 years until Census Day (April 1, 2020), there is little room for delay. Therefore, to ensure that emerging plans related to the hard-to-count efforts integrate with existing plans, Bureau management will need to continue its focus on control of the changes in hard-to-count efforts moving forward.

- Second, the Bureau faces a challenge of a tighter labor market than existed prior to 2010 that could potentially create shortfalls or delays in its hiring of partnership staff who are needed to reach small and hard-tocount communities. In early hiring for 2020, Bureau officials reported smaller than expected applicant pools, declined offers, and turnover. Although it has plans to identify critical skills for 2020 and for tailored recruiting, collecting data on its hiring efforts will also be important. Currently, the Bureau lacks data from its 2010 Census that could have helped inform its partnership-staff hiring efforts for 2020.

July 26, 2018

The Honorable Claire McCaskill
Ranking Member
Committee on Homeland Security and Governmental Affairs
United States Senate

The Honorable Gary Peters
Ranking Member
Subcommittee on Federal Spending Oversight
and Emergency Management
Committee on Homeland Security and Governmental Affairs
United States Senate

The Honorable Elijah E. Cummings
Ranking Member
Committee on Oversight and Government Reform
House of Representatives

A goal of the 2020 Census is to count everyone once, only once, and in the right place. However, achieving a complete and accurate census is becoming an increasingly complex task. This is in part because the nation's population is growing larger; more diverse in culture, living arrangements, and in the number of languages spoken; and more reluctant to participate. The U.S. Census Bureau's (Bureau) efforts during the early part of the decade were largely focused on preparing and testing innovations largely designed to save money. The former Bureau director previously testified that part of these savings could go toward improving the enumeration of the "hard-to-count" groups historically missed in the census, such as racial and ethnic minorities, renters, and young children. However, the Bureau's cost estimates have risen over the decade, eliminating much of those initial projected savings, and the Bureau has requested increased funding over multiple years while it develops its plans for enumerating these groups.

You asked us to review the Bureau's plans for enumerating the hard-to-count groups in the 2020 Census. This chapter examines (1) the Bureau's plans for improving the enumeration of the hard-to-count in 2020, and how that compares with its effort for 2010; and (2) the challenges, if any, the Bureau faces in improving the enumeration of the hard-to-count in 2020.

We generally limited our scope to the six Bureau operations with a goal or objective related to improving the enumeration of the hard-to-count groups. These operations are: Integrated Partnership and Communications, Language Services, Non-ID Processing, Group Quarters/Service-Based Enumeration, Enumeration at Transitory Locations, and Coverage Improvement.[1]

---

[1] For the 2020 Census, the Bureau included its Coverage Improvement initiative as part of the Non-Response Follow-up operation.

To address both of these objectives, we reviewed Bureau planning, budget, operational, and evaluation documents related to the Bureau's efforts to enumerate hard-to-count groups in the 2020 Census and prior decennials. We also interviewed Bureau officials responsible for planning and executing the 2020 Census and with experience in prior decennials about planned changes from the 2010 Census and related challenges.

To address the first objective, we also analyzed the Bureau's most recently available planned hiring and life cycle cost estimates for these activities and compared those to the Bureau reported level of planned hiring and actual spending for the 2010 Census. To assess the reliability of these Bureau reported data, we compared the historical information the Bureau provided us for the 2010 Census with information found in the Bureau's 2010 Census evaluations, and interviewed knowledgeable Bureau officials. We determined that the data were sufficiently reliable to describe the Bureau's planned hiring and cost estimates for the 2020 Census compared to similar activities for the 2010 Census, except where noted.

To address the second objective, we also reviewed related Bureau evaluations and recommendations from the 2010 Census and 2020 Census research and testing activities, and from the Bureau's National Advisory Committee and hard-to-count related working groups to identify challenges the Bureau may face in improving the enumeration of the hard to count. We supplemented our review of Bureau documentation by conducting a search of the literature for academic and other publications related to including hard-to-count groups in surveys.

We conducted this performance audit from November 2017 to July 2018 in accordance with generally accepted government auditing standards. Those standards require that we plan and perform the audit to obtain sufficient, appropriate evidence to provide a reasonable basis for our findings and conclusions based on our audit objectives. We believe that the evidence obtained provides a reasonable basis for our findings and conclusions based on our audit objectives.

BACKGROUND

Hard-to-Count Groups

Although the Bureau goes to great lengths to conduct an accurate count of the nation's population, some degree of inaccuracy is inevitable. When the census misses a person who should have been included, it results in an undercount. An overcount occurs when an individual is counted more than once or in the wrong place. These errors are problematic because certain groups such as minorities, young children, and renters are more likely to be missed in the census, while other groups such as those who may own a second, seasonal home are more likely to be counted more than once. As census data are used to apportion seats in Congress, redraw congressional districts, and allocate billions of dollars in federal assistance each year, improving coverage and reducing undercounts are important.

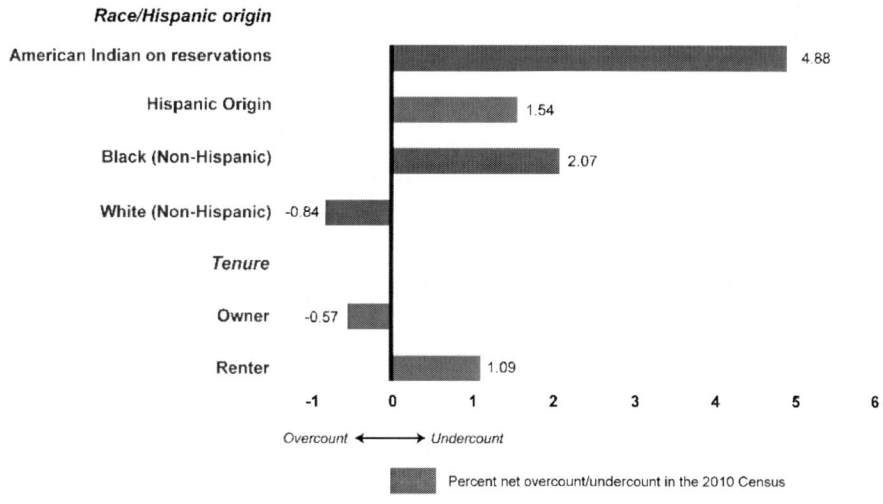

Source: GAO analysis of Census Bureau documentation. | GAO-18-599.
Note: Census Bureau reported that each of the percent net undercounts included were significantly different from zero at the 90 percent confidence level.

Figure 1. Certain Sociodemographic Groups Experienced Differential Undercounts in the 2010 Census.

Table 1. Census Bureau recognizes a range of sociodemographic and other groups as hard-to-count

Complex households including those with blended families, multi-generations, or non-relatives
Cultural and linguistic minorities
Displaced persons affected by a disaster
Lesbian gay bisexual transgender queer/questioning persons
Low income persons
Persons experiencing homelessness
Persons less likely to use the Internet and others without Internet access
Persons residing in places difficult for enumerators to access, such as buildings with strict doormen, gated communities, and basement apartments
Persons residing in rural or geographically isolated areas
Persons who do not live in traditional housing
Persons who do not speak English fluently (or have limited English proficiency)
Persons who have distrust in the government
Persons with mental and/or physical disabilities
Persons without a high school diploma
Racial and ethnic minorities
Renters
Undocumented immigrants (or recent immigrants)
Young children
Young, mobile persons

Source: GAO analysis of Census Bureau documentation. | GAO-18-599.

As an example, the Bureau reported that the 2010 Census did not have a significant net undercount or overcount nationally. However, as shown in figure 1, errors in census coverage were unevenly distributed through the population. For example, the Bureau estimated that it missed nearly 5 percent of American Indians living on reservations—the sociodemographic group with the highest percent net undercount in 2010—whereas the Bureau estimated it overcounted almost 1 percent of non-Hispanic whites.

In addition to those groups with characteristics the Bureau can measure—based on their responses to certain questions asked on the census questionnaire—there are many other hard-to-count groups, some of which cut across sociodemographic groups, as shown in table 1. For example, lesbian, gay, bisexual, transgender, or queer/questioning persons or persons who distrust government can cut across all sociodemographic groups.

There are complex reasons why certain groups are considered hard-to-count. According to Bureau officials, for example, one way to think about the hard-to-count problem is to consider what groups are hard to locate, contact, persuade, and interview for the census (see figure 2).[2]

Hard-to-Locate

Some groups are hard-to-locate because where they live is unknown, or they move frequently. For example, the Bureau faces difficulty counting persons experiencing homelessness. Adding to this difficulty are reported increases in the prevalence and complexity of outdoor encampments across the country. Inhabitants design many of these encampments to remain hidden; some people may remain in an encampment for years while other people may move frequently.

Hard-to-Persuade

Other groups are hard-to-persuade to participate in the census. For example, while the Bureau had identified those who distrust government as a hard-to-count group based on research prior to the 2010 Census, in November 2017, the Bureau reported to its National Advisory Committee an increase in unprompted confidentiality concerns raised by individuals in focus groups and pretests for the 2020 Census and other surveys.[3]

---

[2] Bureau officials said one way they think about "hard to count" is an organizing principle, based on research contained in Roger Tourangeau, Brad Edwards, Timothy Johnson, Kirk Wolter, and Nancy Bates, *Hard to Survey Populations* (Cambridge: Cambridge University Press, 2014).

[3] U.S. Census Bureau, Center for Survey Measurement, *Respondent Confidentiality Concerns*, Memorandum for Associate Directorate for Research and Methodology (September 20, 2017).

Source: GAO analysis. | GAO-18-599.

Figure 2. Difficulty Locating, Contacting, Persuading, and Interviewing Different Hard-to-Count Groups Illustrate the Complexity of Achieving a Complete Census.

Multiple Factors

Some groups are hard to count for multiple complex reasons. For example, a Bureau taskforce found that households with young children up to 4 years old may be missed altogether due to frequent moves between rental units (hard-to-contact). Moreover, some households studied—such as complex households with multiple generations—also appeared to be confused about whether or not to include their young children when completing the questionnaire or when being interviewed by census enumerators. The Bureau also found that language barriers sometimes resulted in households leaving young children off their census or other survey questionnaire (hard-to-interview).[4]

Congress Has Prioritized Funding for Decennial Partnership and Communications Efforts in Both the 2010 Census and the 2020 Census

An appropriation in the American Recovery and Reinvestment Act of 2009 (Recovery Act) allowed the Bureau to increase the funding of the Bureau's 2010 Census partnership and communications efforts.[5] The Bureau has partnered with governments, businesses, and local community organizations to help promote the census. The Bureau has also relied on a communications campaign including paid advertisements in national and targeted markets to help build awareness of the census. After adjusting for inflation, the Bureau spent about $123 million to expand its advertising and about $125 million to expand its partnership efforts (in 2017 dollars), primarily by hiring additional partnership-related staff beyond original plans.

Partnership staff hired to support the 2010 Census were responsible for mobilizing local support for the census by working with local organizations to promote participation. Partnership staff for the 2010

---

[4] The Bureau created a task force to research the undercount of young children in 2013. The task force has issued a number of reports on different facets of the undercount problem. For example, in 2014, the Bureau released its first report on the topic. U.S. Census Bureau, *The Undercount of Young Children* (February 2014). The Bureau has released additional related reports, including one on child undercount probes in the 2010 Census. U.S. Census Bureau, *Decennial Statistics Studies Division, Investigating the 2010 Undercount of Young Children – Child Undercount Probes* (January 2017).

[5] Pub. L. No. 111-5, div. A, tit. II. The Bureau received $1 billion from the Recovery Act. In the conference report accompanying the act, the conferees stated that "of the amounts provided, up to $250,000,000 shall be for partnership and outreach efforts to minority communities and hard-to-reach populations." H.R. Conf. Rep. No. 116-16 at 417 (2009). All figures in then-year dollars as reported in the act. The purpose of including information related to the Recovery Act is to provide contextual information for when comparing the Bureau's plans from the 2010 and 2020 Censuses.

Census included a mix of partnership specialists—responsible for building relationships with and obtaining commitments from governments, local businesses, and other organizations to help promote the census—managers, graphic designers, and clerical support positions. After receiving Recovery Act funding, the Bureau created a new partnership assistant position. After the partnership specialists had established agreements with local organizations, these partnership assistants were responsible for supporting the implementation of promotion efforts, such as by staffing fairs and other events. Bureau officials told us that they believed that creating a new partnership assistant position would help promote census awareness.

The Consolidated Appropriations Act, 2018 directed the Bureau to conduct its fiscal year 2018 partnership and communications efforts in preparation for the 2020 Census at a level and staffing no less than the Bureau conducted during fiscal year 2008 in preparation for the 2010 Census. The act appropriated more than $2.5 billion for the Periodic Censuses and Programs account, which according to Bureau officials includes over $1 billion from the Bureau's fiscal year 2019 budget request intended to smooth transition of funding between fiscal years, such as in the event of a continuing resolution.[6]

THE BUREAU PLANS TO ENHANCE OUTREACH TO AND THE ENUMERATION OF HARD-TO-COUNT GROUPS IN 2020, BUT ESTIMATED SPENDING IS SIMILAR TO 2010

The Bureau Plans Enhancements to Key 2020 Census Operations to Address Complexity Enumerating Hard-to-Count Groups

The Bureau will continue to rely on its Integrated Partnership and Communications operation—designed to communicate the importance of

---

[6] Consolidated Appropriations Act, 2018, Pub. L. No. 115-141, div. B, tit. I.

census participation and motivate self-response—as a key component of its efforts to improve enumeration of hard-to-count persons in the 2020 Census. Evaluations conducted by the Bureau found that its partnership and communications efforts had positive effects on increasing awareness and participation among the hard-to-count in prior censuses. Because of the positive effects, the Bureau has begun outreach to the more than 257,000 tribal, state, and local governments as well as other businesses and organizations it partnered with in 2010. For example, the Bureau plans to continue using "trusted voices"—individuals or groups with relevance, importance, and relatability to a given population, such as local leaders and gatekeepers within isolated communities—to promote the census. As part of this effort, the Bureau plans to continue outreach initiatives to specific constituencies, such as to faith-based communities, and, through its Foreign Born/Immigrant initiative, to outreach and communicate with recent immigrants, undocumented residents, refugees, and migrant and seasonal farm workers.

In addition, the Bureau still plans to advertise in national and targeted markets. For example, to support its 2020 outreach efforts, including to hard-to-count groups, the Bureau awarded a communications contract in August 2016 to Young and Rubicam, an advertising firm. As has been done in prior censuses, this contractor has enlisted 14 partners and subcontractors to help it reach specific sociodemographic groups, such as American Indian and Alaska Native populations and Hispanic communities.[7]

Given the increasingly complex task of counting those historically missed in the census, the Bureau has taken steps or plans to enhance some aspects of the initiatives under its Integrated Partnership and Communications operation and to other key operations compared to the

---

[7] The Bureau publicly released the first of three planned versions of its Integrated Communications Plan in October 2017, outlining the high-level components of the communications campaign being developed by Young and Rubicam in consultation with the Bureau. U.S. Census Bureau, *2020 Census Integrated Communications Plan Version 1.0* (June 2, 2017).

2010 Census, as shown in table 2.[8] For example, the Bureau overhauled a metric it has used to help manage and target field work for its partnerships to areas with hard-to-count populations, basing it now on predictions of each household's likelihood to self-respond to the census.[9] Using this new low response score metric, the Bureau created a publicly available online mapping tool for its partnership staff and other users to better understand the sociodemographic make-up of their assigned areas and to plan their outreach efforts accordingly.[10] Moreover, as we previously recommended in 2010, the Bureau also plans to develop predictive models to help allocate its advertising using: (1) these predictive response data, (2) results describing the complexity of difficult enumeration from its recent "behaviors, attitudes, and motivators survey" study and focus groups, and (3) other third-party data.[11]

The Bureau is still evaluating certain initiatives before deciding whether or not to include them in its 2020 plans. For example, as part of the 2018 End-to-End Test currently underway in Providence, Rhode Island, the Bureau is piloting the use of Internet kiosks in selected post offices to help allow persons to self-respond to the census. Bureau officials said they will decide whether to move forward with the use of kiosks in post offices in 2020 after evaluating the pilot and the test.

---

[8] We examined the Bureau's plans for the 2020 Census in the six Bureau operations that we believed were most directly related to the hard-to-count. Within the Non-Response Follow-up operation, we focused on the Bureau's Coverage Improvement initiative.

[9] The new "low response score" metric uses tract and block-level housing, demographic, and socio-demographic variables from the 2010 Census, five-year American Community Survey estimates, and other 2010 Census data. This low response score replaces the Bureau's prior "hard-to-count" score used for the 2010 Census that was based on mail-back response information from the 2000 Census, including socio-demographic variables as well as the number of vacant homes and the percentage of people living below the poverty line in each given census tract or geographic area.

[10] The Bureau has made its Response Outreach Area Mapper (ROAM) application available at www.census.gov/roam.

[11] GAO, 2010 Census: Key Efforts to Include Hard-to-Count Populations Went Generally as Planned; Improvements Could Make the Efforts More Effective for Next Census, GAO-11-45 (Washington, D.C.: Dec. 14, 2010).

Table 2. Examples of planned changes to 2020 Census Hard-to-Count Operations from 2010 to help address increasing complexities

Operation	Description	Planned change for 2020 Census
Integrated Partnership and Communications	Communicate the importance of participating in the census by engaging and motivating people to self respond	• Enhance use of data to support predictive modeling • Use focus groups with hard-to-count groups to help develop communications message, rather than just to test it • Publish public tool to map low-response areas • Consult with American Indian and Alaskan • Native tribes (2 years earlier than for last census) • Formally promote state and local complete count committees to improve the count (one year earlier) • Expand college outreach to community colleges, vocational schools, and off-campus housing
Coverage Improvement[a]	Resolve errors with people who were counted in the wrong place, counted more than once, or missed	• Revise enumerator job aides and probes on questionnaires to emphasize counting young children
Group Quarters/Service Based Enumeration	Enumerate people in group quarters (colleges, nursing homes, etc.) and people experiencing homelessness when receiving services at soup kitchens, mobile food vans, and targeted outdoor locations	• Allow administrators of target programs to transfer enumeration data electronically

Table 2. (Continued)

Operation	Description	Planned change for 2020 Census
Language Support	Support language needs of non-English speaking populations through non-English language response options and guidance materials	• Increase number of non-English language response options and instructional materials • Formalize in-house language translation expertise
Non-ID Processing	Verify addresses for responses that do not already include a unique Census ID on them	• Allow for near real-time validation of Internet responses • Incorporate administrative records and other improvements to improve acceptance rates
Enumeration at Transitory Locations	Enumerate people at transitory locations who do not have a usual home elsewhere, such as recreational vehicle parks, campgrounds, racetracks, carnivals, marinas, hotels, and motels	• Changes not yet known; detailed operational plan scheduled for release in September 2018

Source: GAO analysis of Census Bureau documentation. | GAO-18-599.
[a] The Bureau included its Coverage Improvement initiative as part of the Non-Response Follow-up operation for the 2020 Census.

In addition, according to the Bureau's current planning documents, the Bureau has plans to change other key operations to help improve the enumeration of certain hard-to-count groups. For example, to help address the complex undercount of young children, the Bureau revised the census questionnaire and instructions to enumerators to more explicitly mention the inclusion of grandchildren and any non-relatives in household population counts. In addition, the Bureau's planning documents describe plans to offer administrators at certain group quarters locations, such as college dormitories, the option to electronically transfer their rosters to the

Bureau.[12] Bureau officials said that this planned change will help reduce the need for enumerators to visit those locations, and that such an efficiency gain will allow them to devote resources on the ground to other harder to enumerate group quarters.[13]

Recognizing the importance of reaching an increasingly linguistically diverse population, the Bureau has also made significant changes to its Language Services operation for 2020, including increasing the number of non-English languages formally supported by the Bureau. Table 3 below summarizes changes in the number of languages the Bureau plans to support. According to the Bureau, this larger choice of languages should increase the percentage of limited-English-speaking households directly supported by that operation from 78 percent in 2010 to 87 percent in 2020.

The Bureau is still assessing the level of non-English support it will directly provide through advertising, partnership, and promotional materials. Bureau officials stated that they will decide the number of — and which—non-English languages to support after it has completed research on how best to segment advertising markets in fall 2018. Until then, it has committed to at least 12 non-English languages—which is less than half of the 27 non-English languages similarly supported in the 2010 Census.[14] Bureau officials said that one action they will take to mitigate any effects if the Bureau decides on a fewer number of languages for 2020 is to provide language-independent media templates—including scripts to

---

[12] The other response option for group quarters will be enumerators visiting these locations and using paper questionnaires to capture responses. The Bureau had initially planned to use mobile devices to enumerate individuals at these locations, including at service-based locations such as shelters and soup kitchens, but in September 2017, announced its decision to revert back to paper-based methods as had been done in prior censuses.

[13] We previously recommended in 2010 that the Bureau determine the factors led to staffing issues we observed during enumeration at such service-based locations and take steps to ensure more efficient staffing levels (GAO-11-45). The Bureau agreed with the recommendation and has taken initial steps to update its staffing model. However, as of June 2018, the Bureau had not provided information to show it had determined the underlying factors that led to the observed overstaffing in order to help prevent a repeat in 2020. We continue to monitor the Bureau's progress in fully implementing the recommendation.

[14] The Bureau provided additional non-English language support in 2010 using Recovery Act funding. According to Bureau officials, prior to the additional Recovery Act funding received in 2009, the Bureau planned to support 13 non-English languages.

videos ready for non-English voiceovers— to any partner groups that may need them.

Table 3. Census Bureau generally plans to expand coverage of non-English language support for the 2020 Census, compared to the 2010 Census

	Language Support	2010 Census (# non-English languages)	Planned 2020 Census (# non-English languages)
Response Options	Paper questionnaire and related mailings	Spanish only	Spanish only
	Internet questionnaire	-	12
	Telephone assistance	5	12
Support Materials	Enumerator glossary to translate technical terms	-	59
	Enumerator job-aid to help identify respondent's preferred language	50	59
	Guides to help respondents complete questionnaire	59	59
	Advertising, partnership, and promotional material	27	Decision pending (at least 12)

Source: GAO analysis of Census Bureau documentation. | GAO-18-599.

The Bureau has also formalized its language translation capabilities for the non-English languages it chooses to support based on 2010 Census evaluations that found, among other things, that the Bureau's lack of sufficient oversight of its translation process hampered consistency of its translation of promotion and outreach materials. For the 2020 Census, Bureau officials said they intend to rely on in-house translation experts adhering to translation industry standards. Bureau officials stated that the Bureau will not attempt to oversee the translations that partners may make into less commonly spoken languages using the Bureau's language-neutral materials when trying to reach more isolated language areas, though officials stated that its partners, including contractors for advertising, will rely on Bureau-developed language glossaries for census terminology when translating into other languages.

The Bureau Plans Total Spending for Its 2020 Census Outreach Efforts Similar to that for 2010 Census, and to Hire More Partnership Specialists Instead of Assistants

The Bureau estimates total spending for its 2020 partnership and communications outreach efforts to be similar to what it reported spending on those efforts for the 2010 Census after adjusting for inflation. Specifically, according to documents supporting the Bureau's most recent life cycle cost estimate for the 2020 Census, the Bureau may spend about $850 million in its outreach to promote the 2020 Census, compared to nearly $830 million in total spending in comparable categories for the 2010 Census.[15] (See table 4.)

Partnership Staff

According to the Bureau's current planning documents, it will hire nearly twice as many partnership specialists— responsible for building relationships and obtaining commitments from organizations—to support the 2020 Census than it hired to support the 2010 Census. Despite this planned increase in partnership specialists, the Bureau's total estimated spending on partnership staff—$248 million—is less than the $334 million

---

[15] All figures are in constant 2017 dollars adjusted for inflation using the fiscal year Gross Domestic Price Index.

the Bureau reported spending in the same cost category for 2010 after adjusting for inflation.[16] This change is in part because the Bureau does not plan to hire any partnership assistants to support the 2020 Census.

Table 4. Census Bureau's 2020 life cycle cost estimates show shifts in spending categories for outreach efforts, compared to the 2010 Census

Dollars in millions		
Categories	2010 actual[a]	2020 estimated
Partnership staff in local and regional offices	334	248
Support from census headquarters	106	111
Communications campaign	388	492
Total	828	851

Source: GAO analysis of Census Bureau documentation. | GAO-18-599.

Note: All data are in constant fiscal year 2017 dollars adjusted for inflation using the fiscal year Gross Domestic Price index.

[a] The Bureau's spending for the 2010 Census includes additional funding ($125 million for partnership and $123 million for communications in 2017 dollars) it received in 2009 pursuant to the American Recovery and Reinvestment Act.

According to Bureau planning data from the 2010 Census, the Bureau planned to hire over 1,700 partnership assistants—those that assisted specialists for the 2010 Census—with Recovery Act funding. As noted previously, Bureau officials said that the additional funding it received from the Recovery Act in 2009 (about $125 million in 2017 dollars) largely funded the hiring of these partnership assistants. The effect of the Recovery Act funding on partnership hiring is shown in figure 3 below. According to Bureau officials, without the Recovery Act funding and its direction for the Bureau to increase hiring in order to stimulate the economy, the Bureau would not have hired the large number of partnership assistants that it did.

According to Bureau officials, this shift in hiring toward more partnership specialists will enable a greater focus on creating more

---

[16] Without the Recovery Act funding, according to Bureau documentation, the Bureau's spending on partnership staff in field offices would have been about $209 million in the 2010 Census, about $40 million less than planned for the 2020 Census after adjusting for inflation.

partnerships and require greater reliance on partner organizations to help with staffing for outreach and promotion events in local communities that partnership assistants were used for in the 2010 Census. While the ability of future partners to help with these events remains to be seen, Bureau officials involved in early outreach with partners stated that they believe this planned approach shows early promise based on the over 1,500 partners they have engaged for the 2020 Census so far.

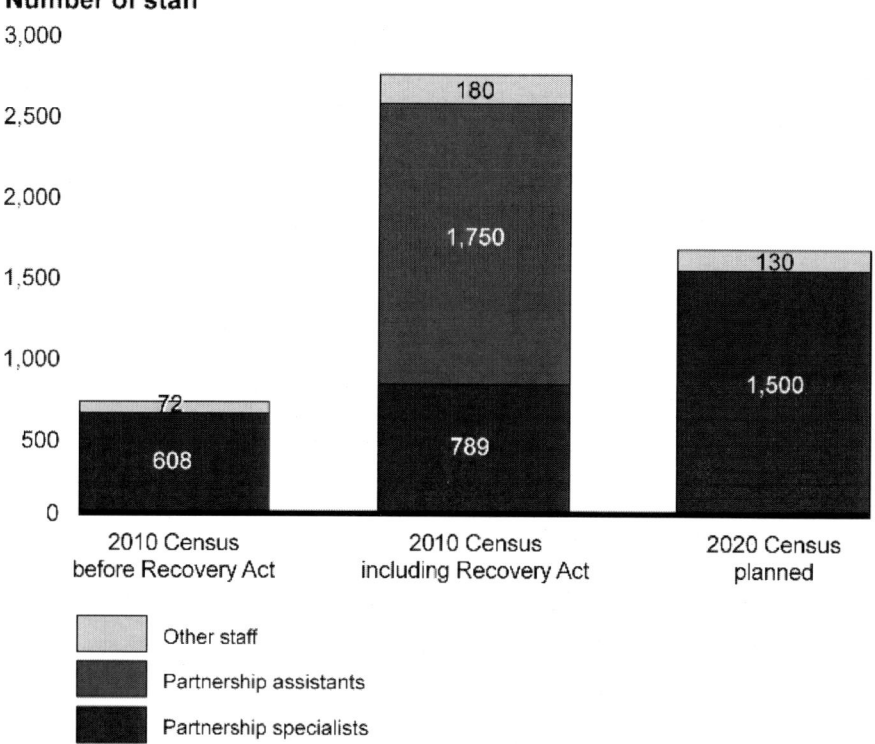

Source: GAO analyses of Census Bureau documentation. | GAO-18-599.

Figure 3. Census Bureau Plans to Hire More Partnership Specialists and No Partnership Assistants for 2020 Census, Compared to the 2010 Census Plans.

Headquarters Support

The $111 million amount the Bureau plans to spend in headquarters support for outreach efforts is similar to the $106 million it spent in the

2010 Census after adjusting for inflation. According to Bureau documents, this support will be used for advertising, media, and partnership efforts.

Communications Campaign

The Bureau plans to spend more in its communications campaign category in the 2020 Census than what it reported spending in this area during the 2010 Census—$492 million compared to $388 million after adjusting for inflation, according to the Bureau's cost estimation documents. The campaign will include paid advertising and the development of promotional materials. According to Bureau officials, they will initiate much of this spending in May 2019. This larger figure includes about $152 million for additional contracted services still being planned, but provisionally allocated for various advertising support efforts with the balance for various partnership materials not included in other contracts.

The Bureau does not plan to repeat its "2010 Census Road Tour" involving a large mobile display and over a dozen cargo vans that were driven to promotional events around the country at a cost of about $16.6 million after adjusting for inflation. While the Bureau did not conduct a formal evaluation of the initiative's effectiveness at encouraging response during the 2010 Census, Bureau officials told us that they do not believe it was as effective a use of resources compared to the other options they are planning for 2020.

The Bureau Started Partnership Hiring Earlier for the 2020 Census Than for 2010

An evaluation conducted by the Bureau of its 2010 partnership efforts recommended that, for the 2020 Census, the Bureau hire at least a core group of partnership staff 3 years prior to census day instead of the 2 years prior as was done for the 2010 Census. Consistent with that recommendation, according to Bureau officials, the Bureau hired five partnership specialists for the 2020 Census in October 2015—more than 2 years earlier in the decennial cycle than its first hiring of partnership specialists in January 2008 for the 2010 Census, as shown in figure 4. Bureau officials told us that this hiring helped the Bureau complete tribal consultations earlier than it had for the 2010 Census. Moreover, the Bureau continued its early hiring with 39 more partnership specialists in fiscal year 2017. Bureau officials said that, with the additional year of preparation, these staff initiated outreach to the highest level of government in each of the 50 states, the District of Columbia, and Puerto Rico, resulting in, as of April 2018, partnership staff having obtained commitments or statements of interest from all but two state governments to form State Complete Count Commissions/Committees. These Commissions/Committees are intended to help form partnerships at the highest levels of government within each state and leverage each state's vested interest in a timely and complete count of its population.

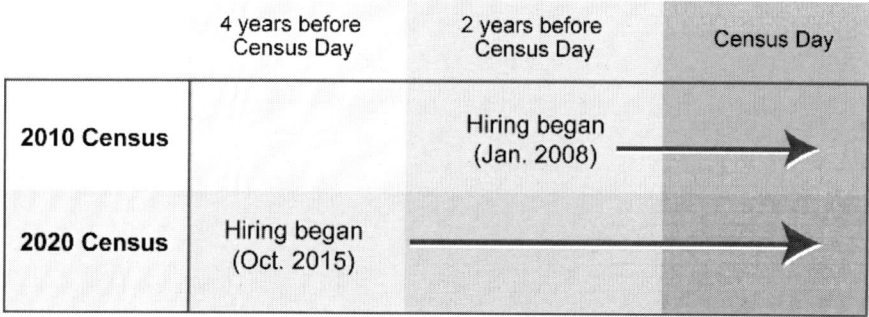

Source: GAO analyses of Census Bureau documentation. | GAO-18-599.

Figure 4. Census Bureau Began Hiring 2020 Partnership Staff 2 Years Earlier Than for 2010 Census.

Bureau officials said they also recently further accelerated the Bureau's planned time frames for hiring partnership specialists. These officials said that with the funds made available in the Fiscal Year 2018 Consolidated Appropriations Omnibus, the Bureau began posting job announcements for about 70 partnership specialists in April 2018 and hopes to begin hiring in July 2018—3 months earlier than October 2018, as had otherwise been planned.

In addition, with 2018 funds, Bureau officials said they are working to identify elements of the communications campaign to begin earlier than the planned start date of October 1, 2018. The Bureau and the lead communications contractor identified possible efforts to start months earlier. According to Bureau officials, they are finalizing how to accelerate these efforts, including the *Statistics in Schools* initiative, media planning, and hosting a creative development workshop with the communications contractors.[17]

THE BUREAU FACES INTERNAL MANAGEMENT AND EXTERNAL WORKFORCE CHALLENGES IN IMPROVING THE ENUMERATION OF HARD-TO-COUNT GROUPS IN 2020

The Bureau Faces an Internal Management Challenge Integrating Its Many Hard-to-Count Efforts

According to the Bureau and as shown in figure 5, over one-third of its 35 operations (14 of 35) are designed, at least in part, to help improve the enumeration of hard-to-count groups. These efforts range from the earliest field data collection operations—such as *address canvassing* when the Bureau aims to identify all possible addresses where people live, including hidden housing units such as basement apartments or attics—to some of its later field operations, such as *nonresponse follow-up* when census enumerators visit each household that did not self-respond.

Each of the 35 operations is implemented by a separate team that manages and controls its activities and, according to Bureau governance documents, is also responsible for reviewing and managing its risks, schedule, and scope, as well as developing needed capability requirements. Team leads are responsible for ensuring integration with other operation teams, and escalating risks to management, as well as ensuring

---

[17] The Statistics in Schools initiative is intended to promote census participation and use of census data by providing resources and learning activities to teachers, students, and parents.

communication upward to the various governance bodies overseeing the decentralized structure.[18] Operational decisions within the scope of plans that have been approved by the governance bodies are made at the team level, while ultimate responsibility rests with respective associate directors for the decennial, field, communications, and other directorates, whose staff largely comprise the teams, and the Director of the Census Bureau itself. The Bureau exercises change control over the scope, schedule, and documentation of its baseline program design, with a change control board comprising process and program managers with responsibility over the operational teams. Approved changes are formally communicated via e-mail to stakeholders in the change control process.

2020 Census Operations

SUPPORT								
Program Management	Census/Survey Engineering				Infrastructure			
Program Management	Security, Engineering & Integration	Security, Privacy, and Confidentiality	Content and Form Design	Language Services	Decennial Service Center	Field Infrastructure	Decennial Logistics Management	IT Infrastructure

FRAME	RESPONSE DATA			PUBLISH DATA
Geographic Programs	Forms Printing and Distribution	Non-ID Processing	Nonresponse Followup	Data Products and Dissemination
	Paper Data Capture	Update Enumerate	Response Processing	Redistricting Data Program
Local Update of Census Addressing	Integrated Partnership and Communications	Group Quarters	Federally Affiliated Count Overseas	Count Review
Address Canvassing		Enumeration at Transitory Locations		Count Question Resolution
	Internet Self-Response	Census Questionnaire Assistance	Update Leave	Archiving

OTHER CENSUSES	TEST AND EVALUATION			
Island Areas Censuses	Coverage Measurement Design & Estimation	Coverage Measurement Matching	Coverage Measurement Field Operations	Evaluations and Experiments

Source: GAO analyses of Census Bureau documentation. | GAO-18-599.
Note: Gray shading indicates an operation included in the draft Census Bureau Plan for Enumerating the Hard-to-Count.

Figure 5. Over One-Third of 2020 Census Operations Involve Efforts Intended to Improve Enumeration of the Hard-to-Count Groups.

Managing decentralized operations in such a way can be effective and provide an agency flexibility in responding to changing conditions on the

---

[18] Decentralization is an organizational structure in which operations and decision-making responsibilities are delegated by top management to others within the organization.

ground, such as when adapting census methods in response to natural disasters as the Bureau had to do during the 2010 Census for areas affected by Hurricane Katrina. However, such decentralization also presents a challenge to management as it tries to ensure the integration of its efforts to improve enumeration of the hard-to-count groups.

To help address the challenge of managing so many hard-to-count efforts that cut across the decentralized operations, during our review, the Bureau developed a draft operational design document. This document describes the major operations and initiatives that contribute to, at least in part, its goal to improve the enumeration of hard-to-count groups in the 2020 Census. This is the Bureau's first comprehensive look at the hard-to-count goal for the 2020 Census. Bureau officials said that they developed the document because they realized that looking across the Bureau's operations and how they relate to difficulties enumerating hard-to-count groups would provide them a useful perspective that could help identify any gaps or interdependencies in their various hard-to-count efforts. Bureau officials said they plan to refine and include this document as a chapter in the fall 2018 update of their broader 2020 Census Operational Plan. Although this is a good first step to elevate the visibility of the hard-to-count goal, we identified a number of other areas where additional steps or management focus may be needed in order to help ensure integration of certain hard-to-count related efforts, including the following:

- During exchanges of information between the Bureau and its National Advisory Committee in 2017 and 2018, the Bureau proposed using additional focus groups with certain population groups, census interviewers, and trusted community messengers. These focus groups are intended to identify root causes and ways to overcome the confidentiality concerns increasingly being raised by respondents in the Bureau's earlier testing by helping to inform messaging and outreach plans as well as staff support documents and training materials. However, as of May 1, 2018, the Bureau reported that it had yet to identity the resources needed to conduct the additional focus groups it had proposed. If the Bureau is going to take this step, it would need to complete its analysis from these proposed focus groups with interviewers and others before starting to develop its 2020 messaging, currently scheduled to begin in October 2018. Any delays in scheduling these activities could have an effect on activities intended to help improve enumeration of the hard-to-count in other related operations.[19]

- The detailed operational plans for 10 of the Bureau's 14 hard-to-count-related operations have been documented and released publicly. However, we found that several of the detailed plans already released—while self-described as being updated over time to reflect changes in strategies based on ongoing planning, research, and testing—are nearly two years old and may not reflect more recent decisions made. Attention by Bureau management to the details of these operational plans as they are updated will be critical to ensure that their interdependencies with other efforts are accounted for.

- Similarly, as of May 2018, little detail is available about what interdependencies the other 4 hard-to-count related efforts will have on the overall 2020 Census Operational Plan and on the Bureau's efforts to improve the enumeration of the hard-to-count in particular. For example, the Bureau's operation to enumerate persons at transitory locations—key to counting mobile persons, including those living at motels or with traveling carnivals—is one of the 4 efforts without a detailed operational plan yet. Because the Bureau is not scheduled to test the integration of this enumeration with other systems before the 2020 Census, it remains to be seen how its forthcoming design may interact with other related operations and systems. While Bureau officials stated that procedures likely to be used for this operation are well established from prior censuses, they also stated that there may be significant

---

[19] In technical comments in response to a draft of this chapter, Bureau officials told us that they do not have plans to conduct the additional proposed focus groups with census interviewers and trusted community messengers.

changes from the past in the process the Bureau uses to determine where to count persons in this operation and may rely on changes in the non-ID processing operation—helping enumerate persons not having a pre-assigned census identification number. With less than 2 years to go until Census Day (April 1, 2020), there is little room for delay in considering how forthcoming details on hard-to-count efforts yet to be finalized— or changed based on ongoing testing or other decisions—may have consequences on other related efforts.

According to the Project Management Institute's *A Guide to the Project Management Body of Knowledge*, integrated change control can help address overall risk to related efforts, which often arises from changes made without consideration of the overall goals or plans.[20] A significant amount of hard-to-count-related planning for the 2020 Census is currently underway, and in the less than 2 years remaining before Census Day, it will be important for Bureau management to maintain a focus that helps ensure that hard-to-count-related decisions yet to be made as well as any changes to those already made are integrated with other related efforts. Focused attention on these efforts will also help ensure that any interdependencies, synergies, or gaps are identified and included in the change-control processes the Bureau already has in place.

Hiring Partnership Staff with Critical Skills in a Tight Labor Market Creates a Workforce Challenge for the Bureau and It Lacks Data from 2010 to Guide Its Efforts

As noted previously, a key component of the 2010 Census was the hiring of partnership staff to help build relationships with and obtain commitments from local organizations to help encourage census

---

[20] Project Management Institute, Inc., *A Guide to the Project Management Body of Knowledge (PMBOK® Guide)*, Sixth Edition (Newton Square, PA: 2017). PMBOK is a trademark of Project Management Institute, Inc.

participation, particularly among hard-to-count groups. For the 2020 Census, in addition to the core relationship-building skills, Bureau officials said they are working to identify specialized skills needed to operate partnership initiatives in a 2020 environment, such as advanced knowledge of digital media. However, the Bureau faces a significant challenge in hiring these kinds of staff because it is operating in a much tighter labor market than it did prior to the 2010 Census. As a result, it may not be able to hire the partnership staff with the skills it now needs as easily as it had in the past.

According to Bureau of Labor Statistics data, the unemployment rate in January 2008, when the Bureau first hired partnership staff for the 2010 Census, was 5 percent. That number increased to more than 7 percent by December 2008, and then ranged from more than 7.5 percent to 10 percent in 2009 and through Census Day in April 2010. During this time, the Bureau hired nearly 3,000 partnership staff, many of which the Bureau hired in a few short months after receiving additional funding from the Recovery Act. The unemployment rate is substantially lower now as we approach the comparable part of the decade for the 2020 Census. Specifically, the rate has ranged from 4.9 percent in October 2016, when the Bureau starting hiring for an early round of about 40 partnership staff, to less than 4 percent in May 2018.

Bureau officials reported experiencing challenges during these early hiring efforts for partnership staff, although they were ultimately able to fill the nearly 40 positions the Bureau sought to fill across its six census regions. Bureau officials in the regional field offices reported observing smaller applicant pools, declined job offers, and early turnover due to a lower pay rate the Bureau offered compared to the local economy. Moreover, these officials reported seeing fewer applicants through local job markets, which had been successful recruiting mechanisms in the prior census. According to the Bureau's planning documents, the Bureau plans to ramp up its hiring of partnership specialists between July 2018 and 2019. If the unemployment rate generally holds steady at around the 4 percent of May 2018, the Bureau will likely face challenges recruiting and retaining partnership staff with the critical skills needed.

Bureau officials said that they will develop customized recruiting strategies to fill specific needs as they identify and refine the mix of partnership skills needed to support their 2020 efforts. For example, Bureau officials acknowledged the need to more effectively use USAJobs, the federal recruiting website, and targeted job announcements. They also identified the possibility of hiring additional partnership staff for short-term assignments closer to census day to help meet specific needs, such as assisting with non-English language enumeration and connecting with faith-based or immigrant communities in areas with low participation. Following through on its plans to identify an optimal mix of skill-sets and tailored recruiting strategies, in accordance with leading practices, will be important for the Bureau as it operates in a tight labor market because delays or shortfalls in hiring partnership staff could put the Bureau's plans for building support for the census at risk.[21]

As the Bureau has decided to rely more heavily on partnership specialists as part of its outreach and promotion strategy to reach hard-to-count groups and still faces decisions about where to staff them, it has done so without the benefit of data on its actual hiring of partnership staff from the 2010 Census. During our review, the Bureau was unable to readily provide us with data on the actual number or timing of partnership specialists and assistants hired to support the 2010 Census, and instead, we had to use detailed Bureau planning documents for our analysis. Bureau officials reported that their records in 2010 did not clearly link the positions and grades recorded in the payroll system for individual staff who were hired to support a different operation to the roles they subsequently played in carrying out the partnership efforts.

Standards for Internal Control in the Federal Government state that management should use quality information to achieve the entity's objectives.[22] Bureau officials recognize the importance of having such data readily available both for evaluating implemented efforts and for future

---

[21] GAO, Key Issues, Best Practices and Leading Practices in Human Capital Management; and Human Capital: Effective Use of Flexibilities Can Assist Agencies in Managing Their Workforces, GAO-03-2 (Washington, D.C.: Dec. 6, 2002).

[22] GAO, Standards for Internal Control in the Federal Government, GAO-14-704G (Washington, D.C.: September 2014).

planning, and said they will take steps to better record these types of data for the 2020 Census. Doing so will better position the Bureau to evaluate the effectiveness of its hiring strategy and tradeoffs in alternative approaches, to learn lessons from the 2020 implementation, and to optimize related staffing strategies in the future.

CONCLUSION

Much of the Bureau's planning efforts to help address the longstanding challenge of enumerating hard-to-count groups in the 2020 Census are underway. Importantly, the various operations and initiatives related to the hard-to-count are either in the planning or early implementation stages. While the Bureau has taken some steps to better understand the scope of these efforts, going forward, it will be important for the Bureau to ensure that management maintains a focus on forthcoming changes and decisions on hard-to-count related efforts to ensure they are integrated with other hard-to-count related efforts across the Bureau's decentralized operations. Doing so will help the Bureau identify possible synergies, interdependencies, or gaps specific to how they might affect the Bureau's ability to improve the census and help address overall risk to related efforts.

In addition, information about related efforts in prior censuses can help inform management and its ongoing planning. However, the Bureau's lack of complete and reliable data on hiring partnership staff for the 2010 Census—such as numbers, dates, and positions filled—affects its ability to fully consider tradeoffs it is making among types of staff it plans to hire for the 2020 Census. As the Bureau continues to ramp up its hiring of partnership specialists and other staff to support enumeration of the hard-to-count, improved recording of hiring numbers, dates, and positions filled—particularly for staff supporting multiple operations—can help position the Bureau to evaluate the effectiveness of its hiring strategy and support efforts to optimize any related hiring in future censuses.

RECOMMENDATIONS FOR EXECUTIVE ACTION

We are making the following two recommendations to the Department of Commerce and the Census Bureau:
- The Secretary of Commerce should ensure the Director of the U.S. Census Bureau takes steps to ensure that forthcoming changes and decisions on hard-to-count related efforts are integrated with other hard-to-count related efforts across the Bureau's decentralized operations. (Recommendation 1)
- The Secretary of Commerce should ensure the Director of the U.S. Census Bureau takes steps to ensure for the purposes of evaluation and future planning that information is recorded and available on partnership hiring numbers, dates, positions filled, and in support of what part of the 2020 Census. (Recommendation 2)

AGENCY COMMENTS AND OUR EVALUATION

We provided a draft of this chapter to the Department of Commerce. In its written comments, reproduced in appendix I the Department of Commerce agreed with our findings and recommendations and said it would develop an action plan to address them. The Census Bureau also provided technical comments that we incorporated, as appropriate.

As agreed with your offices, unless you publicly announce the contents of this chapter earlier, we plan no further distribution until 30 days from the report date. At that time, we will send copies of this chapter to the Secretary of Commerce, the Undersecretary of Economic Affairs, the Acting Director of the U.S. Census Bureau, and the appropriate congressional committees.

Robert Goldenkoff
Director, Strategic Issues

Appendix I: Comments from the Department of Commerce

UNITED STATES DEPARTMENT OF COMMERCE
The Secretary of Commerce
Washington, D.C. 20230

July 18, 2018

Mr. Robert Goldenkoff
Director
Strategic Issues
United States Government Accountability Office
Washington, DC 20548

Dear Mr. Goldenkoff:

The U.S. Department of Commerce appreciates the opportunity to comment on the United States Government Accountability Office's (GAO) draft report: *"2020 Census: Actions Needed to Address Challenges to Enumerating Hard-to-Count Groups"* (GAO-18-599).

The Department agrees with the findings and recommendations in this draft report. Once GAO issues the final report, we will prepare an action plan to document the steps we will take regarding the final recommendations.

Sincerely,

Wilbur Ross

In: Actions Needed ...
Editor: Bryant Schneider

ISBN: 978-1-53616-716-0
© 2019 Nova Science Publishers, Inc.

Chapter 2

2020 CENSUS: ACTIONS NEEDED TO ADDRESS KEY RISKS TO A SUCCESSFUL ENUMERATION[*]

Robert Goldenkoff and Nick Marinos

WHY GAO DID THIS STUDY

The Bureau is responsible for conducting a complete and accurate decennial census of the U.S. population. The decennial census is mandated by the Constitution and provides vital data for the nation. A complete count of the nation's population is an enormous undertaking as the Bureau seeks to control the cost of the census, implement operational innovations, and use new and modified IT systems. In recent years, GAO has identified challenges that raise serious concerns about the Bureau's ability to conduct a cost-effective count. For these reasons, GAO added the 2020 Census to its High-Risk list in February 2017.

---

[*] This is an edited, reformatted and augmented version of United States Government Accountability Office; Testimony Before the Committee on Homeland Security and Governmental Affairs, U.S. Senate, Publication No. GAO-19-588T, dated July 16, 2019.

GAO was asked to testify about the reasons the 2020 Census remains on the High-Risk List and the steps the Bureau needs to take to mitigate risks to a successful census. To do so, GAO summarized its prior work regarding the Bureau's planning efforts for the 2020 Census. GAO also included preliminary observations from its ongoing work examining the IT systems readiness and cybersecurity for the 2020 Census. This information is related to, among other things, the Bureau's progress in developing and testing key systems and the status of cybersecurity risks.

WHAT GAO RECOMMENDS

Over the past decade, GAO has made 106 recommendations specific to the 2020 Census to help address issues raised in this and other products. The Department of Commerce has generally agreed with the recommendations. As of June 2019, 31 of the recommendations had not been fully implemented.

WHAT GAO FOUND

The 2020 Decennial Census is on GAO's list of high-risk programs primarily because the Department of Commerce's Census Bureau (Bureau) (1) is using innovations that are not expected to be fully tested, (2) continues to face challenges in implementing information technology (IT) systems, and (3) faces significant cybersecurity risks to its systems and data. Although the Bureau has taken initial steps to address risk, additional actions are needed as these risks could adversely impact the cost, quality, schedule, and security of the enumeration.

- Innovations. The Bureau is planning several innovations for the 2020 Census, including allowing the public to respond using the internet. These innovations show promise for controlling costs, but they also introduce new risks, in part, because they have not been used extensively, if at all, in earlier enumerations. As a result, testing is essential to ensure that key IT systems and operations will function as planned. However, citing budgetary uncertainties, the Bureau scaled back operational tests in 2017 and 2018, missing an opportunity to fully demonstrate that the innovations and IT systems will function as intended during the 2020 Census. To manage risk to the census, the Bureau has developed hundreds of mitigation and contingency plans. To maximize readiness for the 2020 Census, it will also be important for the Bureau to prioritize among its mitigation and contingency strategies those that will deliver the most cost-effective outcomes for the census.

- Implementing IT systems. The Bureau plans to rely heavily on IT for the 2020 Census, including a total of 52 new and legacy IT systems and the infrastructure supporting them. To help improve its implementation of IT, in October 2018, the Bureau revised its systems development and testing schedule to reflect, among other things, lessons learned during its 2018 operational test. However, GAO's ongoing work has determined that the Bureau is at risk of not meeting near-term IT system development and testing schedule milestones for five upcoming 2020 Census operational deliveries, including self-response (e.g., the ability to respond to the 2020 Census through the internet). These schedule management challenges may compress the time available for the remaining system development and testing, and increase the risk that systems will not function as intended. It will be important that the Bureau effectively manages IT implementation risk to ensure that it meets near-term milestones for system development and testing, and that it is ready for the major operations of the 2020 Census.
- Cybersecurity. The Bureau has established a risk management framework that requires it to conduct a full security assessment for nearly all the systems expected to be used for the 2020 Census and, if deficiencies are identified to determine the corrective actions needed to remediate those deficiencies. As of the end of May 2019,

the Bureau had over 330 corrective actions from its security assessments that needed to be addressed, including 217 that were considered "high-risk" or "very high-risk." However, of these 217 corrective actions, the Bureau identified 104 as being delayed. Further, 74 of the 104 were delayed by 60 or more days. According to the Bureau, these corrective actions were delayed due to technical challenges or resource constraints. GAO recently recommended that the Bureau take steps to ensure that identified corrective actions for cybersecurity weaknesses are implemented within prescribed time frames. Resolving identified vulnerabilities more timely can help reduce the risk that unauthorized individuals may exploit weaknesses to gain access to sensitive information and systems.

To its credit, the Bureau is also working with the Department of Homeland Security (DHS) to support its 2020 Census cybersecurity efforts. For example, DHS is helping the Bureau ensure a scalable and secure network connection for the 2020 Census respondents and to strengthen its response to potential cyber threats. During the last 2 years, as a result of these activities, the Bureau has received 42 recommendations from DHS to improve its cybersecurity posture. GAO recently recommended that the Bureau implement a formal process for tracking and executing appropriate corrective actions to remediate cybersecurity findings identified by DHS. Implementing the recommendation would help better ensure that DHS's efforts result in improvements to the Bureau's cybersecurity posture.

In addition to addressing risks which could affect innovations and the security of the enumeration, the Bureau has the opportunity to improve its cost estimating process for the 2020 Census, and ultimately the reliability of the estimate itself, by reflecting best practices. In October 2017, the 2020 Census life-cycle cost estimate was updated and is now projected to be $15.6 billion, a more than $3 billion (27 percent) increase over its earlier estimate. GAO reported in August 2018 that although the Bureau had taken steps to improve its cost estimation process for 2020, it needed to implement a system to track and report variances between actual and estimated cost elements. According to Bureau officials, they planned to release an updated version of the 2020 Census life-cycle estimate in the spring of 2019; however, they had not done so as of June 28, 2019. To ensure that future updates to the life-cycle cost estimate reflect best practices, it will be important for the Bureau to implement GAO's recommendation related to the cost estimate.

Over the past decade, GAO has made 106 recommendations specific to the 2020 Census to help address these risks and other concerns. The Department of Commerce has generally agreed with these recommendations and has taken action to address many of them. However, as of June 2019, 31 of the recommendations had not been fully implemented. While all 31 open recommendations are important for a high-quality and cost-effective enumeration, 9 are directed at managing the risks introduced by the Bureau's planned innovations for the 2020 Census. To ensure a high-quality and cost-effective enumeration, it will be important for the Bureau to address these recommendations.

Chairman Johnson, Ranking Member Peters, and Members of the Committee:

We are pleased to be here today to discuss the U.S. Census Bureau's (Bureau) progress in preparing for the 2020 Decennial Census. Conducting the decennial census of the U.S. population is mandated by the Constitution and provides vital data for the nation. The information that the census collects is used to apportion the seats of the House of Representatives; redraw congressional districts; allocate billions of dollars each year in federal financial assistance; and provide a social, demographic, and economic profile of the nation's people to guide policy decisions at each level of government. Further, businesses use census data to market new services and products and to tailor existing ones to demographic changes.

A complete count of the nation's population is an enormous undertaking. The Bureau, a component of the Department of Commerce (Commerce), is seeking to control the cost of the 2020 Census while it implements several innovations and manages the processes of acquiring and developing information technology (IT) systems.

In recent years, we have identified challenges that raise serious concerns about the Bureau's ability to conduct a cost-effective count of the nation, including issues with the agency's research, testing, planning, scheduling, cost estimation, systems development, risk management, and cybersecurity practices.

Over the past decade, we have made 106 recommendations specific to the 2020 Census to help address these and other concerns. Commerce has generally agreed with our recommendations and has made progress in implementing them. However, 31 of the recommendations had not been fully implemented as of June 2019, although the Bureau had taken initial steps to address many of them. In addition, one recommendation was closed as the Bureau decided to implement a different approach than the one about which the recommendation was directed.

We added the 2020 Decennial Census to our high-risk list in February 2017, and it remains on our high-risk list today.[1] As preparations for the next census continue to ramp up, fully implementing our recommendations to address the risks jeopardizing the 2020 Census is more critical than ever.

At your request, our testimony today will describe (1) why the 2020 Decennial Census remains a high-risk area and (2) the steps that Commerce and the Bureau need to take going forward to mitigate the risks jeopardizing a secure and cost-effective census.

The information in this statement is based primarily on our prior work regarding the Bureau's planning efforts for 2020.[2] For that body of work,

---

[1] GAO, *High-Risk Series: Substantial Efforts Needed to Achieve Greater Progress on High-Risk Areas,* GAO-19-157SP (Washington, D.C.: Mar. 6, 2019) and *High-Risk Series: Progress on Many High-Risk Areas, While Substantial Efforts Needed on Others,* GAO-17-317 (Washington, D.C.: Feb. 15, 2017). GAO maintains a high-risk program to focus attention on government operations that it identifies as high-risk due to their greater vulnerabilities to fraud, waste, abuse, and mismanagement or the need for transformation to address economy, efficiency, or effectiveness challenges.

[2] For example, GAO, *2020 Census: Additional Actions Needed to Manage Risk,* GAO-19-399 (Washington, D.C.: May 31, 2019); *2020 Census: Additional Steps Needed to Finalize Readiness for Peak Field Operations,* GAO-19-140 (Washington, D.C.: Dec. 10, 2018); *2020 Census: Continued Management Attention Needed to Address Challenges and Risks with Developing, Testing, and Securing IT Systems,* GAO-18-655 (Washington, D.C.: Aug. 30, 2018); *2020 Census: Bureau Has Made Progress with Its Scheduling, but*

we reviewed, among other things, relevant Bureau documentation, including the 2020 Census Operational Plan; recent decisions on preparations for the 2020 Census; and outcomes of key IT milestone reviews.

In the summer of 2018 we visited the Bureau's 2018 End-to-End test site in Providence County, Rhode Island to observe door-to-door field enumeration during the non-response follow-up, an operation where enumerators personally visit each non-responding household to include them in the census. We also discussed the status of our recommendations with Commerce and Bureau staff. Other details on the scope and methodology for our prior work are provided in each published report on which this testimony is based.

In addition, we included information in this statement from our ongoing work on the readiness of the Bureau's IT systems for the 2020 Census. Specifically, we collected and reviewed documentation on the status and plans for system development and testing, and for addressing cybersecurity risk, for the 2020 Census. This includes the Bureau's integration and implementation plan, memorandums documenting outcomes of security assessments, and reports prepared by the Department of Homeland Security (DHS) for the Bureau on cybersecurity risks. We also interviewed relevant agency officials.

We provided a copy of the applicable new information that we are reporting in this testimony to the Bureau and DHS for comment on June 25, 2019. The Bureau provided technical comments, which we addressed as appropriate.

We conducted the work on which this statement is based in accordance with generally accepted government auditing standards. Those standards require that we plan and perform the audit to obtain sufficient, appropriate evidence to provide a reasonable basis for our findings and conclusions based on our audit objectives. We believe that the evidence obtained

---

Further Improvement Will Help Inform Management Decisions, GAO-18-589 (Washington, D.C.: July 26, 2018); and, *2020 Census: Actions Needed to Address Challenges to Enumerating Hard-to-Count Groups*, GAO-18-599 (Washington, D.C.: July 26, 2018).

provides a reasonable basis for our findings and conclusions based on our audit objectives.

BACKGROUND

As shown in Table 1 the cost of counting the nation's population has been escalating with each decade. The 2010 Census was the most expensive in U.S. history at about $12.3 billion, and was about 31 percent more costly than the $9.4 billion 2000 Census (in 2020 dollars).[3] According to the Bureau, the total cost of the 2020 Census in October 2015 was estimated at $12.3 billion and in October 2017 that cost estimate grew to approximately $15.6 billion, approximately a $3 billion increase.[4]

Additionally, Bureau officials told us that while the estimated cost of the census had increased to $15.6 billion, it was nevertheless managing the 2020 Census to a lower cost of $14.1 billion. Bureau officials explained that the $14.1 billion includes all program costs and contingency funds to cover risks and general estimating uncertainty. The remaining $1.5 billion estimated cost is additional contingency for "unknown unknowns"—that is, low probability events that could cause massive disruptions—and several what-if scenarios such as an increase in the wage rate or additional supervisors needed to manage field operations.[5]

---

[3] According to the Bureau, these figures rely on fiscal year 2020 constant dollar factors derived from the Chained Price Index from "Gross Domestic Product and Deflators Used in the Historical Tables: 1940–2020" table from the Fiscal Year 2016 Budget of the United States Government.

[4] The historical life-cycle cost figures for prior decennials as well as the initial estimate for 2020 provided by Commerce in October 2017 differ slightly from those reported by the Bureau previously. According to Commerce documents, the more recently reported figures are "inflated to the current 2020 Census time frame (fiscal years 2012 to 2023)," rather than to 2020 constant dollars as the earlier figures had been. Specifically, since October 2017, Commerce and the Bureau have reported the October 2015 estimate for the 2020 Census as $12.3 billion; this is slightly different than the $12.5 billion the Bureau had initially reported.

[5] The $15.6 billion cost estimate for the 2020 Census includes a total of $2.6 billion in contingency funds.

Table 1. The Cost of Previous Decennial Censuses and the Estimated Cost of the 202

Benchmark	Cost	Explanation
2000 Census	$9.4 billion[a]	Final cost of the 2000 Census
2010 Census	$12.3 billion[a]	Final cost of the 2010 Census
2020 Census estimated cost in October 2015	$12.3 billion[a]	Initial cost estimate of the 2020 Census
2020 Census estimated cost in October 2017	$15.6 billion[b]	Revised cost estimate of the 2020 Census
2020 Census cost estimate less a portion of contingency funds	$14.1 billion[b]	Cost estimate the Bureau is managing operations to for the 2020 Census

Source: GAO analysis of Census Bureau data. | GAO-19-588T
Notes:
[a]Constant 2020 dollars.
[b]Inflated to the current 2020 Census time frame, fiscal years 2012 to 2023.

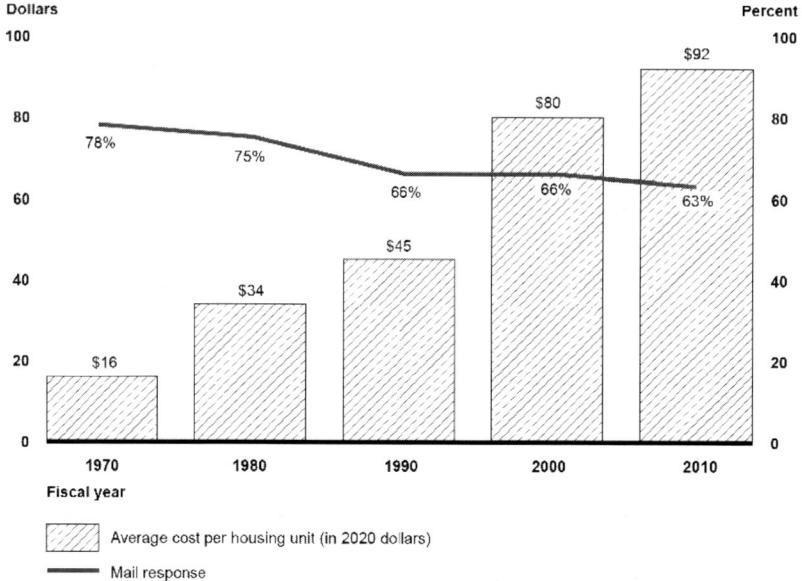

Source: GAO analysis of Census Bureau data. | GAO-19-558T.

Figure 1. The Average Cost of Counting Each Housing Unit (in 2020 Dollars) Has Escalated Each Decade, While the Percentage of Mail Response Rates Has Declined.

Moreover, as shown in Figure 1, the average cost for counting a housing unit increased from about $16 in 1970 to around $92 in 2010 (in 2020 constant dollars). At the same time, the return of census questionnaires by mail (the primary mode of data collection) declined over this period from 78 percent in 1970 to 63 percent in 2010. Declining mail response rates have led to higher costs because the Bureau sends temporary workers to each non-responding household to obtain census data.

Achieving a complete and accurate census has become an increasingly daunting task, in part, because the population is growing larger, more diverse, and more reluctant to participate in the enumeration. In many ways, the Bureau has had to invest substantially more resources each decade to conduct the enumeration.

In addition to these external societal challenges that make achieving a complete count a daunting task, the Bureau also faces a number of internal management challenges that affect its capacity and readiness to conduct a cost-effective enumeration. Some of these issues—such as acquiring and developing IT systems and preparing reliable cost estimates—are long-standing in nature.

At the same time, as the Bureau looks toward 2020, it has faced emerging and evolving uncertainties. For example, on March 26, 2018, the Secretary of Commerce announced his decision to add a question to the decennial census on citizenship status which resulted in various legislative actions[6] and legal challenges.[7] Ultimately, the case was heard by the U.S. Supreme Court, which, in a June 26, 2019, ruling, prevented the addition of the question because the Court found that the evidence Commerce provided in the case did not match the Secretary's explanation.[8] In

---

[6] Commerce, Justice, Science, and Related Agencies Appropriations Act, 2020, H.R. 3055, § 534, 116th Cong. (passed June 25, 2019). Ensuring Full Participation in the Census Act of 2019, H.R. 1734, 116th Cong. (as introduced March 13, 2019); 2020 Census Accountability Act, H.R. 5292, 115th Cong. (as introduced March 15, 2018).

[7] See, e.g., *New York v. U.S. Dept. of Commerce*, No. 18-cv-2921, (S.D.N.Y. Jan. 15, 2019); *California v. Ross*, No. 18-cv-01865, (N.D. Cal. Mar. 6, 2019).

[8] *New York v. U.S. Dept. of Commerce*, U.S. No. 18-966 at *33 (2019). A majority of the Court held in favor of the government on whether the question was permitted under the Enumeration Clause of the Constitution and the Census Act, but remanded to the Southern District Court of New York for additional proceedings on the limited question of whether the administrative record demonstrated reasonable decision making.

addition, the Fourth Circuit Court of Appeals remanded other legal challenges to the district court on June 24, 2019, for further legal action, which is yet to be resolved.[9]

According to Bureau officials, on June 28, 2019, Commerce asked the Bureau to put its scheduled July 1 start date for printing questionnaires on hold while it considered legal implications of the Supreme Court ruling. On July 2, 2019, Commerce told the Bureau to proceed with printing questionnaires and other materials without the citizenship question on them. As of July 5, 2019, the Department of Justice (DOJ) indicated that, although printing was continuing without the citizenship question, DOJ was evaluating legal options to include the question.[10]

On July 11, 2019, the President announced that instead of collecting this information from the census questionnaire, he ordered all federal agencies to provide data on citizenship status to Commerce using legally available federal records. We have not analyzed this decision or its implications, if any, for how the Bureau will tabulate its official counts. We will continue to monitor developments for Congress.

The Bureau also faced budgetary uncertainties that, according to the Bureau, led to the curtailment of testing in 2017 and 2018. However, the Consolidated Appropriations Act, 2018 appropriated for the Periodic Censuses and Programs account $2.544 billion, which more than doubled the Bureau's request in the President's Fiscal Year 2018 Budget of $1.251 billion.[11] According to the explanatory statement accompanying the act, the appropriation, which is available through fiscal year 2020, was provided to ensure the Bureau has the necessary resources to immediately address any

---

[9] *La Union Del Pueblo Entero et al. v. Ross*; *Kravitz v. Department of Commerce*, No. 19-1382 (4th Cir.). These cases were remanded to the district court for evidence-gathering on the plaintiffs' equal protection claims.

[10] *Id.* DOJ, on behalf of Commerce, submitted a filing for this case on July 5, 2019 stating its intent to evaluate legal options for including the citizenship question.

[11] Consolidated Appropriations Act, 2018, Pub. L. No. 115-141, Division B, Title I (Mar. 23, 2018). Of the total appropriated for the Periodic Censuses and Programs account, $2.095 billion was for the 2020 Census and $213.6 million was for the American Community Survey.

issues discovered during operational testing, and to provide a smoother transition between fiscal year 2018 and fiscal year 2019.[12]

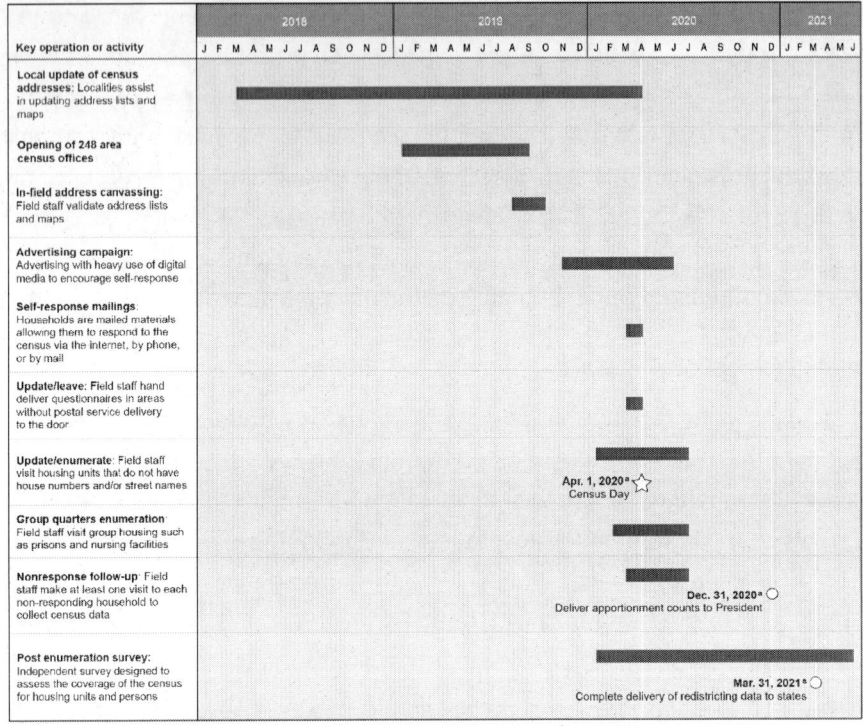

Source: GAO summary of Census Bureau Information. | GAO-19-588T.
[a]Indicates dates that are mandated by law.

Figure 2. Timeline of Selected Decennial Events.

The availability of those resources enabled the Bureau to continue preparations for the 2020 Census during the 35 days in December 2018 to January 2019 when appropriations lapsed for the Bureau and a number of other federal agencies. Moreover, the Consolidated Appropriations Act, 2019 appropriated for the Periodic Censuses and Programs account $3.551

---

[12] Joint explanatory statement of conference, 164 Cong. Rec. H2045, H2084 (daily ed. Mar. 22, 2018) (statement of Chairman Frelinghuysen), specifically referenced in section 4 of the Consolidated Appropriations Act, 2018, Pub. L. No. 115-141, § 4 (Mar. 23, 2018).

billion.[13] According to Bureau officials, this level of funding for fiscal year 2019 is sufficient to carry out 2020 Census activities as planned.

Importantly, the census is conducted against a backdrop of immutable deadlines. In order to meet the statutory deadline for completing the enumeration, census activities need to take place at specific times and in the proper sequence.[14] Thus, it is absolutely critical for the Bureau to stay on schedule. Figure 2 shows some dates for selected decennial events.

The Bureau Has Begun Opening Offices and Hiring Temporary Staff

The Bureau has begun to open its area census offices (ACO) for the 2020 Census. It has signed leases for all 248 ACOs, of which 39 of the offices will be open for the address canvassing operation set to begin in August 2019 where staff verifies the location of selected housing units. The remaining 209 offices will begin opening this fall. In 2010 the Bureau opened 494 census offices. The Bureau has been able to reduce its infrastructure because it is relying on automation to assign work and to record payroll. Therefore there is less paper—field assignments, maps, and daily payroll forms—to manually process.

For the 2020 Census, the Bureau is refining its recruiting and hiring goals, but tentatively plans to recruit approximately 2.24 million applicants and to hire over 400,000 temporary field staff from that applicant pool for two key operations: address canvassing, and nonresponse follow-up, where they visit households that do not return census forms to collect data in person. In 2010 the Bureau recruited 3.8 million applicants and hired 628,000 temporary workers to conduct the address canvassing and nonresponse follow-up field operations. According to Bureau officials, it has reduced the number of temporary staff it needs to hire because automation has made field operations more efficient and there is less paper.

---

[13] Consolidated Appropriations Act, 2019, Pub. L. No. 116-6, Division C, Title I (Feb. 15, 2019). Of the total appropriated for the Periodic Censuses and Programs account, $3.015 billion was for the 2020 Census and $211.4 million was for the American Community Survey.
[14] 13 U.S.C. § 141(b).

As of June 2019, the Bureau reported that for all 2020 Census operations it had processed about 430,000 applicants.

In addition, the Bureau was seeking to hire approximately 1,500 partnership specialists by the end of June 2019 to help increase census awareness and participation in minority communities and hard-to-reach populations. As of July 9, 2019, the Bureau's latest biweekly reporting indicated that it had hired 813 partnership specialists as of June 22, 2019. Moreover, as of July 10, 2019, Bureau officials told us that another 830 applicants were waiting to have their background checks completed. According to Bureau officials, hiring data are based on payroll dates generated biweekly, while background check data are tracked internally. Therefore, according to Bureau officials, more current hiring data were not available as of July 10, 2019 to indicate whether the Bureau had met its June 30 hiring goal.

Among other things, partnership specialists are expected to either provide or identify partners to help provide supplemental language support to respondents locally in over 100 different languages. We will continue to monitor the Bureau's progress in meeting its partnership specialist staffing goals and addressing any turnover that takes place. Hiring partnership specialists in a timely manner and maintaining adequate partnership specialist staffing levels are key to the Bureau's ability to carry out its planned outreach efforts, especially to hard-to-count communities.

Moreover, Bureau officials also stated that the current economic environment (i.e., the low unemployment rate compared to the economic environment of the 2010 Census) has not yet impacted their ability to recruit staff. The Bureau will continue to monitor the impact of low unemployment on its ability to recruit and hire at the local and regional levels.

The Bureau Plans to Rely Heavily on IT for the 2020 Census

For the 2020 Census, the Bureau is substantially changing how it intends to conduct the census, in part by re-engineering key census-taking methods and infrastructure, and making use of new IT applications and systems. For example, the Bureau plans to offer an option for households to respond to the survey via the internet and enable field-based enumerators[15] to use applications on mobile devices to collect survey data from households. To do this, the Bureau plans to utilize 52 new and legacy IT systems, and the infrastructure supporting them, to conduct the 2020 Census.

A majority of these 52 systems have been tested during operational tests in 2017 and 2018. For example, the Bureau conducted its 2018 End-to-End test, which included 44 of the 52 systems and was intended to test all key systems and operations in a census-like environment to ensure readiness for the 2020 Census.

Nevertheless, additional IT development and testing work needs to take place before the 2020 Census. Specifically, officials from the Bureau's Decennial Directorate said they expect that the systems will need to undergo further development and testing due to, among other things, the need to add functionality that was not part of the End-to-End test, scale system performance to support the number of respondents expected during the 2020 Census, and address system defects identified during the 2018 End-to-End test.

To prepare the systems and technology for the 2020 Census, the Bureau is also relying on substantial contractor support. For example, it is relying on contractors to develop a number of systems and components of the IT infrastructure, including the IT platform that is intended to be used to collect data from households responding via the internet and telephone, and for non-response follow-up activities. Contractors are also deploying the IT and telecommunications hardware in the field offices and providing device-as-a-service capabilities by procuring the mobile devices and cellular service to be used for non-response follow-up.[16]

---

[15] Enumerators are Census Bureau employees who travel from door-to-door throughout the country to try to obtain census data from individuals who do not respond through other means, including the internet, on paper, or by phone.

[16] In non-response follow-up, if a household does not respond to the census by a certain date, the Bureau will send out employees to visit the home. The Bureau's plan is for these

In addition to the development of technology, the Bureau is relying on a technical integration contractor to integrate all of the key systems and infrastructure. The contractor's work is expected to include, among other things, evaluating the systems and infrastructure and acquiring the infrastructure (e.g., cloud or data center) to meet the Bureau's scalability and performance needs; integrating all of the systems; and assisting with technical, performance and scalability, and operational testing activities.

2020 Census Identified by GAO as a High-Risk Area

In February 2017, we added the 2020 Decennial Census as a high-risk area needing attention from Congress and the executive branch.[17] This was due to significant risks related to, among other things, innovations never before used in prior enumerations,[18] the acquisition and development of IT systems, and expected escalating costs.

Among other things, we reported that the commitment of top leadership was needed to ensure the Bureau's management, culture, and business practices align with a cost-effective enumeration. We also stressed that the Bureau needed to rigorously test census-taking activities; ensure that scheduling adheres to best practices; improve its ability to manage, develop, and secure its IT systems; and have better oversight and control over its cost estimation process.

Our experience has shown that agencies are most successful at removal from our High-Risk List when leaders give top level attention to the five criteria for removal and Congress takes any needed action. The five criteria for removal that we identified in November 2000 are as follows:[19]

---

enumerators to use a census application, on a mobile device provided by the Bureau, to capture the information given to them by the in-person interviews.

[17] GAO-17-317.

[18] The Bureau has fundamentally re-examined its approach for conducting the 2020 Census to help reduce costs. To do this, the agency plans to use innovations in four broad areas (described later in this statement): re-engineering field operations, using administrative records, verifying addresses in-office, and developing an Internet self-response option.

[19] GAO, *Determining Performance and Accountability Challenges and High Risks*, GAO-01-159SP (Washington, D.C.: Nov. 1, 2000).

- Leadership Commitment. The agency has demonstrated strong commitment and top leadership support.
- Capacity. The agency has the capacity (i.e., people and resources) to resolve the risk(s).
- Action Plan. A corrective action plan exists that defines the root causes and solutions, and that provides for substantially completing corrective measures, including steps necessary to implement solutions we recommended.
- Monitoring. A program has been instituted to monitor and independently validate the effectiveness and sustainability of corrective measures.
- Demonstrated Progress. The agency has demonstrated progress in implementing corrective measures and in resolving the high-risk area.

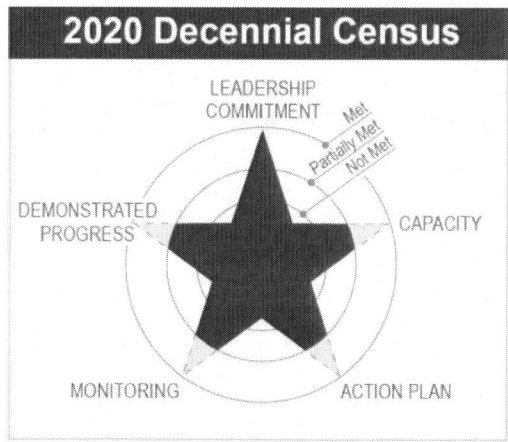

Source: GAO analysis. | GAO-19-588T.

Note: Each point of the star represents one of the five criteria for removal from the High-Risk List and each ring represents one of the three designations: not met, partially met, or met. An unshaded point at the innermost ring means that the criterion has not been met, a partially shaded point at the middle ring means that the criterion has been partially met, and a fully shaded point at the outermost ring means that the criterion has been met.

Figure 3. Status of High-Risk Area for the 2020 Decennial Census, as of March 2019.

These five criteria form a road map for efforts to improve, and ultimately address, high-risk issues. Addressing some of the criteria leads to progress, while satisfying all of the criteria is central to removal from the list. As we reported in the March 2019 high-risk report,[20] the Bureau's efforts to address the risks and challenges for the 2020 Census had fully met one of the five criteria for removal from the High-Risk List—leadership commitment—and partially met the other four, as shown in Figure 3. Additional details about the status of the Bureau's efforts to address this high-risk area are discussed later in this statement.

THE 2020 CENSUS REMAINS HIGH RISK DUE TO CHALLENGES FACING THE ENUMERATION

The 2020 Census is on our list of high-risk programs because, among other things, (1) innovations never before used in prior enumerations are not expected to be fully tested, (2) the Bureau continues to face challenges in implementing IT systems, (3) the Bureau faces significant cybersecurity risks to its systems and data, and (4) the Bureau's cost estimate for the 2020 Census was unreliable.[21] If not sufficiently addressed, these risks could adversely impact the cost and quality of the enumeration. Moreover, the risks are compounded by other factors that contribute to the challenge of conducting a successful census, such as the nation's increasingly diverse population and concerns over personal privacy.

Key Risk #1: The Bureau Redesigned the Census to Control Costs, and Will Need to Take Several Actions to Better Manage Risks

The basic design of the enumeration—mail out and mail back of the census questionnaire with in-person follow-up for non-respondents—has

---

[20] GAO-19-157SP.
[21] GAO-17-317.

been in use since 1970. However, a lesson learned from the 2010 Census and earlier enumerations is that this traditional design is no longer capable of cost-effectively counting the population.

In response to its own assessments, our recommendations, and studies by other organizations, the Bureau has fundamentally re-examined its approach for conducting the 2020 Census. Specifically, its plan for 2020 includes four broad innovation areas: re-engineering field operations, using administrative records, verifying addresses in-office, and developing an internet self-response option (see Table 2).

If they function as planned, the Bureau initially estimated that these innovations could result in savings of over $5 billion (in 2020 constant dollars) when compared to its estimates of the cost for conducting the census with traditional methods. However, in June 2016, we reported that the Bureau's initial life-cycle cost estimate developed in October 2015 was not reliable and did not adequately account for risk.[22]

As discussed earlier in this statement, the Bureau has updated its estimate from $12.3 billion and now estimates a life-cycle cost of $15.6 billion, which would result in a smaller potential savings from the innovative design than the Bureau originally estimated. According to the Bureau, the goal of the cost estimate increase was to ensure quality was fully addressed.

While the planned innovations could help control costs, they also introduce new risks, in part, because they include new procedures and technology that have not been used extensively in earlier decennials, if at all. Our prior work has shown the importance of the Bureau conducting a robust testing program, including the 2018 End-to-End test.[23] Rigorous testing is a critical risk mitigation strategy because it provides information on the feasibility and performance of individual census-taking activities, their potential for achieving desired results, and the extent to which they are able to function together under full operational conditions.

---

[22] GAO, *2020 Census: Census Bureau Needs to Improve Its Life-Cycle Cost Estimating Process*, GAO-16-628 (Washington, D.C.: June 30, 2016).

[23] GAO, *2020 Census: Bureau Needs to Better Leverage Information to Achieve Goals of Reengineered Address Canvassing*, GAO-17-622 (Washington, D.C.: July 20, 2017).

Table 2. The Census Bureau (Bureau) Is Introducing Four Innovation Areas for the 2020 Census

Innovation area	Description
Re-engineered field operations	The Bureau intends to automate data collection methods, including its case management system.
Administrative records	In certain instances, the Bureau plans to reduce enumerator collection of data by using administrative records (information already provided to federal and state governments as they administer other programs such as, Medicare and Medicaid).
Verifying addresses in-office	To ensure the accuracy of its address list, the Bureau intends to use "in-office" procedures and on-screen imagery to verify addresses and reduce street-by-street field canvassing.
Internet self-response option	The Bureau plans to offer households the option of responding to the survey through the internet. The Bureau has not previously offered such an option on a large scale

Source: GAO analysis of Census Bureau data. | GAO-19-588T.

To address some of these challenges we have made numerous recommendations aimed at improving reengineered field operations, using administrative records, verifying the accuracy of the address list, and securing census responses via the internet.

The Bureau has held a series of operational tests since 2012, but according to the Bureau, it scaled back its most recent field tests because of funding uncertainties. For example, the Bureau canceled the field components of the 2017 Census Test including non-response follow-up, a key census operation.[24] In November 2016, we reported that the cancelation of the 2017 Census Test was a lost opportunity to test, refine, and integrate operations and systems, and that it put more pressure on the 2018 End-to-End test to demonstrate that enumeration activities will function under census-like conditions as needed for 2020.

---

[24] In non-response follow-up, if a household does not respond to the census by a certain date, the Bureau will conduct an in-person visit by an enumerator to collect census data using a mobile device provided by the Bureau.

However, in May 2017, the Bureau scaled back the operational scope of the 2018 End-to-End test and, of the three planned test sites, only the Rhode Island site would fully implement the 2018 End-to-End test. The Washington and West Virginia sites would test just one field operation. In addition, due to budgetary concerns, the Bureau delayed ramp up and preparations for its coverage measurement operation (and the technology that supports it) from the scope of the test.[25] However, removal of the coverage measurement operation did not affect testing of the delivery of apportionment or redistricting data.

Without sufficient testing, operational problems can go undiscovered and the opportunity to improve operations will be lost, in part because the 2018 End-to-End test was the last opportunity to demonstrate census technology and procedures across a range of geographic locations, housing types, and demographic groups under decennial-like conditions prior to the 2020 Census.

We reported on the 2018 End-to-End test in December 2018 and noted that the Bureau had made progress addressing prior test implementation issues but still faced challenges.[26] As the Bureau studies the results of its testing to inform the 2020 Census, it will be important that it addresses key program management issues that arose during implementation of the test. Namely, by not aligning the skills, responsibilities, and information flows for the first-line supervisors during field data collection, the Bureau limited its role in support of enumerators within the re-engineered field operation.

The Bureau also lacked mid-operation training or guidance, which, if implemented in a targeted, localized manner, could have further helped enumerators navigate procedural modifications and any commonly encountered problems when enumerating. It will be important for the Bureau to prioritize its mitigation strategies for these implementation issues so that it can maximize readiness for the 2020 Census.

---

[25] Coverage measurement evaluates the quality of the census data by estimating the census coverage based on a post-enumeration survey.
[26] GAO-19-140.

The Bureau Has Developed Hundreds of Risk Mitigation and Contingency Plans, but Those We Reviewed Were Missing Key Information

To manage risk to the 2020 Census the Bureau has developed hundreds of risk mitigation and contingency plans. Mitigation plans detail how an agency will reduce the likelihood of a risk event and its impacts, if it occurs. Contingency plans identify how an agency will reduce or recover from the impact of a risk after it has been realized.

In May 2019, we reported that the Bureau had identified 360 active risks to the 2020 census as of December 2018—meaning the risk event could still occur and adversely impact the census.[27] Of these, 242 met the Bureau's criteria for requiring a mitigation plan and, according to the Bureau's risk registers, 232 had one (see table 3).[28] In addition, 146 risks met the Bureau's criteria for requiring a contingency plan and, according to the Bureau's risk registers, 102 had one.

Table 3. 2020 Census Risks with Required Mitigation and Contingency Plans, as of December 2018

Plan	Risks requiring plan	Risks with plan
Mitigation	242	232 (96%)
Contingency	146	102 (70%)

Source: GAO analysis of U.S. Census Bureau 2020 Census risk registers. | GAO-19-588T.

Bureau guidance states that these plans should be developed as soon as possible after a risk is added to the risk register, but it does not establish a clear time frame for doing so. Consequently, some risks may go without required plans for extended periods. We found that, as of December 2018, some of the risks without required plans had been added to the Bureau's risk registers in recent months, but others had been added more than 3 years earlier.

---

[27] GAO-19-399.

[28] The Bureau's risk registers catalogue information regarding all risks to the 2020 Census that the Bureau has identified, including risk descriptions, and mitigation and contingency plans.

We reviewed the mitigation and contingency plans in detail for six risks which the Bureau identified as among the major concerns that could affect the 2020 Census. These included cybersecurity incidents, late operational design changes, and integration of the 52 systems and 35 operations supporting the 2020 Census.

We found that the plans did not consistently include key information needed to manage the risk. For example, the Bureau's contingency plan for late operational design changes did not include activities specific to the three most likely late operational design changes—including removal of the citizenship question as a result of litigation or congressional action—that the Bureau could carry out to lessen their adverse impact on the enumeration, should they occur.

We found that gaps stemmed from either requirements missing from the Bureau's decennial risk management plan, or that risk owners—the individuals assigned to manage each risk—were not fulfilling all of their risk management responsibilities. Bureau officials said that risk owners are aware of these responsibilities but do not always fulfill them given competing demands.

Bureau officials also said that they are managing risks to the census, even if not always reflected in their mitigation and contingency plans. However, if such actions are reflected in disparate documents or are not documented at all, then decision makers are left without an integrated and comprehensive picture of how the Bureau is managing risks to the census.

We made seven recommendations to improve the Bureau's management of risks to the 2020 Census, including that the Bureau develop mitigation and contingency plans for all risks that require them, establish a clear time frame for plan development, and ensure that the plans have the information needed to manage the risk. Commerce agreed with our recommendations and said it would develop an action plan to address them.

Key Risk #2: The Bureau Faces Challenges in Implementing IT Systems

We have previously reported that the Bureau faces challenges in managing and overseeing IT programs, systems, and contractors supporting the 2020 Census.[29] Specifically, we have noted challenges in the Bureau's efforts to manage, among other things, the schedules and contracts for its systems. As a result of these challenges, the Bureau is at risk of being unable to fully implement the systems necessary to support the 2020 Census and conduct a cost-effective enumeration.

The Bureau Has Made Initial Progress against Its Revised Development and Testing Schedule, but Risks Missing Near-term Milestones

To help improve its implementation of IT for the 2020 Census, the Bureau revised its systems development and testing schedule. Specifically, in October 2018, the Bureau organized the development and testing schedule for its 52 systems into 16 operational deliveries.[30] Each of the 16 operational deliveries has milestone dates for, among other things, development, performance and scalability testing, and system deployment. According to Bureau officials in the Decennial Directorate, the schedule was revised, in part, due to schedule management challenges experienced, and lessons learned, while completing development and testing during the 2018 End-to-End test.

The Bureau has made initial progress in executing work against its revised schedule. For example, the Bureau completed development of the systems in the first operational delivery—for 2020 Census early operations

---

[29] GAO, *2020 Census: Further Actions Needed to Reduce Key Risks to a Successful Enumeration*, GAO-19-431T (Washington, D.C.: Apr. 30, 2019) and GAO-18-655.

[30] The 52 systems being used in the 2020 Census are to be deployed multiple times in a series of operational deliveries (which include operations such as address canvassing or self-response). That is, a system may be deployed for one operation in the 2020 Census (such as address canvassing), and be deployed again for a subsequent operation in the test (such as self-response). As such, additional development and testing may occur each time a system is deployed.

preparations—in July 2018, and deployed these systems into production in October 2018.

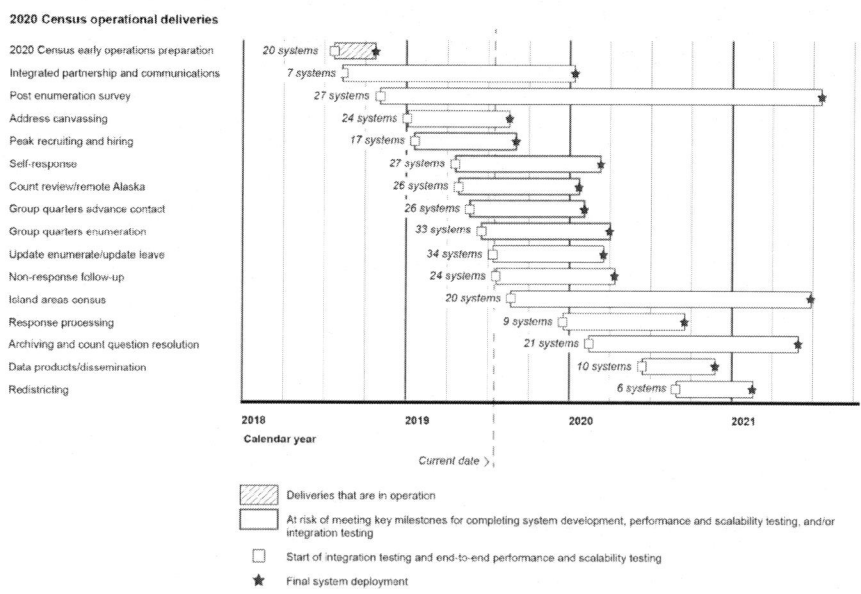

Source: GAO analysis of Census Bureau data. | GAO-19-588T.

Note: The 52 systems being used in the 2020 Census are to be deployed multiple times in a series of operational deliveries (which include operations such as address canvassing or self-response). That is, a system may be deployed for one operation in the 2020 Census (such as address canvassing), and be deployed again for a subsequent operation in the test (such as self-response). As such, additional development and testing may occur each time a system is deployed.

Figure 4. Status of 16 Operational Deliveries for the 2020 Census, as of June 2019.

However, our current work has determined that the Bureau is at risk of not meeting several near-term systems testing milestones. As of June 2019, 11 systems that are expected to be used in a total of five operational deliveries were at risk of not meeting key milestones for completing system development, performance and scalability testing, and/or integration testing.[31] These 11 systems are needed for, among other things,

---

[31] As of June 2019, the 11 systems were 2020 Website; Enterprise Census and Survey Enabling Platform-Operational Control System; Census Questionnaire Assistance; Automated

data collection for operations, business and support automation, and customer support during self-response. Figure 4 presents an overview of the status for all 16 operational deliveries, as of June 2019.

The Bureau Faces Additional Risks Due to Compressed IT Development and Testing Time Frames

The at-risk systems previously discussed add uncertainty to a highly compressed time frame over the next 6 months. Importantly, between July and December 2019, the Bureau is expected to be in the process of integration testing the systems in 12 operational deliveries. Officials from the Bureau's integration contractor noted concern that the current schedule leaves little room for any delays in completing the remaining development and testing activities.

In addition to managing the compressed testing time frames, the Bureau also has to quickly finalize plans related to its IT infrastructure. For example, as of June 2019, the Bureau stated that it was still awaiting final approval for its Trusted Internet Connection.[32] Given that these plans may impact systems being tested this summer or deployed into production for the address canvassing operation in August 2019, it is important that the Bureau quickly addresses this matter.

Our past reporting noted that the Bureau faced significant challenges in managing its schedule for system development and testing that occurred in 2017 and 2018.[33] We reported that, while the Bureau had continued to make progress in developing and testing IT systems for the 2020 Census, it had experienced delays in developing systems to support the 2018 End-to-

---

Tracking and Control; Enterprise Census and Survey Enabling Platform– Internet Self-Response; Census Data Lake; Master Address File/Topologically Integrated Geographic Encoding and Referencing Database; MOJO Field Processing; Production Environment for Administrative Records Staging, Integration, and Storage; Self-Response Quality Assurance; and Unified Tracking System.

[32] External network traffic (traffic that is routed through agency's external connections) must be routed through a Trusted Internet Connection. External connections include those connections between an agency's information system or network and the globally-addressable internet or a remote information system or network and networks located on foreign territory.

[33] GAO-18-655.

End test. These delays compressed the time available for system and integration testing and for security assessments.

In addition, several systems experienced problems during the test. We noted then, and reaffirm now, that continued schedule management challenges may compress the time available for the remaining system and integration testing and increase the risk that systems may not function or be as secure as intended.

The Bureau has acknowledged that it faces risks to the implementation of its systems and technology. As of May 2019, the Bureau had identified 17 high risks related to IT implementation that may have substantial technical and schedule impacts if realized. Taken together, these risks represent a cross-section of issues, such as schedule delays for a fraud-detection system, the effects of late changes to technical requirements, the need to ensure adequate time for system development and performance and scalability testing, contracting issues, privacy risks, and skilled staffing shortages. Going forward, it will be important that the Bureau effectively manages these risks to better ensure that it meets near-term milestones for system development and testing, and is ready for the major operations of the 2020 Census.

Key Risk #3: The Bureau Faces Significant Cybersecurity Risks to Its Systems and Data

The risks to IT systems supporting the federal government and its functions, including conducting the 2020 Census, are increasing as security threats continue to evolve and become more sophisticated. These risks include insider threats from witting or unwitting employees, escalating and emerging threats from around the globe, and the emergence of new and more destructive attacks. Underscoring the importance of this issue, we have designated information security as a government-wide high-risk area since 1997 and, in our most recent biennial report to Congress, ensuring the cybersecurity of the nation was one of nine high-risk areas that we reported needing especially focused executive and congressional attention.[34]

Our prior and ongoing work has identified significant challenges that the Bureau faces in securing systems and data for the 2020 Census.[35] Specifically, the Bureau has faced challenges related to completing security assessments, addressing security weaknesses, resolving cybersecurity recommendations from DHS, and addressing numerous other cybersecurity concerns (such as phishing).[36]

The Bureau Has Made Progress in Completing Security Assessments, but Critical Work Remains

Federal law specifies requirements for protecting federal information and information systems, such as those systems to be used in the 2020 Census. Specifically, the Federal Information Security Management Act of 2002 and the Federal Information Security Modernization Act of 2014 (FISMA) require executive branch agencies to develop, document, and implement an agency-wide program to provide security for the information and information systems that support operations and assets of the agency.[37]

In accordance with FISMA, National Institute of Standards and Technology (NIST) guidance, and Office of Management and Budget (OMB) guidance, the Bureau's Office of the Chief Information Officer (CIO) established a risk management framework. This framework requires system developers to ensure that each of the Bureau's systems undergoes a full security assessment, and that system developers remediate critical deficiencies.

---

[34] GAO-19-157SP.
[35] GAO-19-431T and GAO-18-655.
[36] Phishing is a digital form of social engineering that uses authentic-looking, but fake emails to request information from users or direct them to a fake website that requests information.
[37] The Federal Information Security Modernization Act of 2014, Pub. L. No. 113-283, 128 Stat. 3073 (Dec. 18, 2014) largely superseded the Federal Information Security Management Act of 2002, enacted as Title III, E-Government Act of 2002, Pub. L. No. 107-347, 116 Stat. 2899, 2946 (Dec. 17, 2002).

According to the Bureau's risk management framework, the systems expected to be used to conduct the 2020 Census will need to have complete security documentation (such as system security plans) and an approved authorization to operate prior to their use. As of June 2019, according to the Bureau's Office of the CIO:

- Thirty-seven of the 52 systems have authorization to operate, and will not need to be reauthorized before they are used in the 2020 Census[38]
- Nine of the 52 systems have authorization to operate, and will need to be reauthorized before they are used in the 2020 Census
- Five of the 52 systems do not have authorization to operate, and will need to be authorized before they are used in the 2020 Census
- One of the 52 systems does not need an authorization to operate.[39]

Figure 5 summarizes the authorization to operate status for the systems being used in the 2020 Census, as reported by the Bureau in June 2019.

As we have previously reported, while large-scale technological changes (such as internet self-response) increase the likelihood of efficiency and effectiveness gains, they also introduce many cybersecurity challenges. The 2020 Census also involves collecting personally identifiable information (PII) on over a hundred million households across the country, which further increases the need to properly secure these systems.

---

[38] According to the Bureau's risk management framework, once a system obtains an authorization, it is transitioned to the continuous monitoring process where the authorizing official can provide ongoing authorization for system operation as long as the risk level remains acceptable. Further, according to the framework, authorized systems do not need a formal reauthorization unless the system's authorizing official determines that the risk posture of the system needs to change. This could occur, for example, if the system undergoes significant new development.

[39] According to a June 2019 Bureau memorandum, one system—OneForm Designer Plus—is expected to primarily be used during the 2020 Census as a desktop tool for generating fillable forms. The memorandum further states that, because this system is considered a desktop tool, the Bureau's information security policy does not require it to obtain an authorization to operate.

Thus, it will be important that the Bureau provides adequate time to perform these security assessments, completes them in a timely manner, and ensures that risks are at an acceptable level before the systems are deployed. We have ongoing work examining how the Bureau plans to address both internal and external cyber threats, including its efforts to complete system security assessments and resolve identified weaknesses.

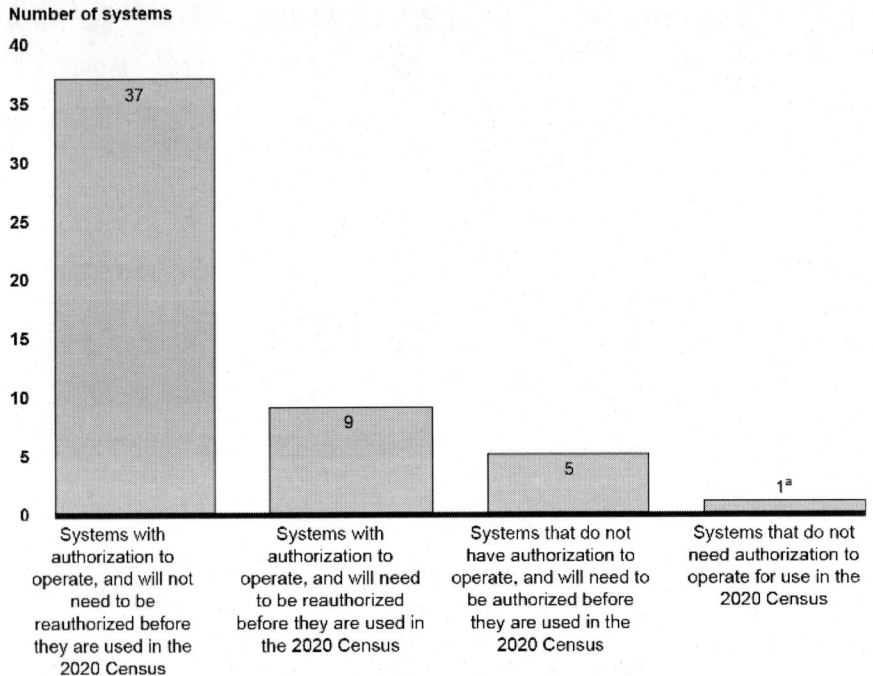

Source: GAO analysis of Census Bureau data. | GAO-19-558T.

[a]According to the Bureau, one system is expected to primarily be used during the 2020 Census as a desktop tool for generating fillable forms and the Bureau's information security policy does not require desktop tools to obtain an authorization to operate.

Figure 5. Authorization to Operate Status for the 52 Systems Being Used by the Census Bureau in the 2020 Census, as Reported by the Bureau in June 2019.

The Bureau Has Identified a Significant Number of Corrective Actions to Address Security Weaknesses, but Has Not Always Been Timely in Completing Them

FISMA requires that agency-wide information security programs include a process for planning, implementing, evaluating, and documenting remedial actions (i.e., corrective actions) to address any deficiencies in the information security policies, procedures, and practices of the agency.[40] Additionally, the Bureau's framework requires it to track security assessment findings that need to be remediated as a plan of action and milestones (POA&M). These POA&Ms are expected to provide a description of the vulnerabilities identified during the security assessment that resulted from a control weakness.

As of the end of May 2019, the Bureau had over 330 open POA&Ms to remediate for issues identified during security assessment activities, including ongoing continuous monitoring. Of these open POA&Ms, 217 (or about 65 percent) were considered "high-risk" or "very high-risk."

While the Bureau established POA&Ms for addressing these identified security control weaknesses, it did not always complete remedial actions in accordance with its established deadlines. For example, of the 217 open "high-risk" or "very high-risk" POA&Ms we reviewed, the Bureau identified 104 as being delayed. Further, 74 of the 104 had missed their scheduled completion dates by 60 or more days. According to the Bureau's Office of Information Security, these POA&Ms were identified as delayed due to technical challenges or resource constraints to remediate and close them.

We previously recommended that the Bureau take steps to ensure that identified corrective actions for cybersecurity weaknesses are implemented within prescribed time frames.[41] As of late May 2019, the Bureau was working to address our recommendation. Until the Bureau resolves identified vulnerabilities in a timely manner, it faces an increased risk, as

---

[40] The Federal Information Security Modernization Act of 2014, Pub. L. No. 113-283, 128 Stat. 3073 (Dec. 18, 2014).
[41] GAO-19-431T.

continuing opportunities exist for unauthorized individuals to exploit these weaknesses and gain access to sensitive information and systems.

The Bureau Is Working with DHS to Improve Its 2020 Census Cybersecurity Efforts, but Lacks a Formal Process to Address DHS's Recommendations

The Bureau is working with federal and industry partners, including DHS, to support the 2020 Census cybersecurity efforts. Specifically, the Bureau is working with DHS to ensure a scalable and secure network connection for the 2020 Census respondents (e.g., virtual Trusted Internet Connection with the cloud), improve its cybersecurity posture (e.g., risk management processes and procedures), and strengthen its response to potential cyber threats (e.g., federal cyber incident coordination).

Federal law describes practices for strengthening cybersecurity by documenting or tracking corrective actions. As previously mentioned, FISMA requires executive branch agencies to establish a process for planning, implementing, evaluating, and documenting remedial actions to address any deficiencies in their information security policies, procedures, and practices. *Standards for Internal Control in the Federal Government* calls for agencies to establish effective internal control monitoring that includes a process to promptly resolve the findings of audits and other reviews.[42] Specifically, agencies should document and complete corrective actions to remediate identified deficiencies on a timely basis. This would include correcting identified deficiencies or demonstrating that the findings and recommendations do not warrant agency action.

Since January 2017, DHS has been providing cybersecurity assistance (including issuing recommendations) to the Bureau in preparation for the 2020 Census. Specifically, DHS has been providing cybersecurity assistance to the Bureau in five areas:

---

[42] GAO, *Standards for Internal Control in the Federal Government*, GAO-14-704G (Washington, D.C.: Sept. 10, 2014).

- management coordination and executive support, including a CyberStat Review;[43]
- cybersecurity threat intelligence and information sharing enhancement through, among other things, a DHS cyber threat briefing to the Bureau's leadership;
- network and infrastructure security and resilience, including National Cybersecurity Protection System (also called EINSTEIN) support;[44]
- incident response and management readiness through a Federal Incident Response Evaluation assessment;[45] and
- risk management and vulnerability assessments for specific high value assets provided by the Bureau.[46]

In the last 2 years, DHS has provided 42 recommendations to assist the Bureau in strengthening its cybersecurity efforts.[47] Among other things, the recommendations pertained to strengthening cyber incident management capabilities, penetration testing[48] and web application assessments of select

---

[43] According to OMB, CyberStat Reviews are face-to-face, evidence-based meetings intended to ensure agencies are accountable for their cybersecurity posture. OMB, DHS, and Commerce participated in the Fiscal Year 2017 CyberStat Review related to the Bureau.

[44] The National Cybersecurity Protection System, operationally known as the EINSTEIN program, is an integrated system-of-systems that is intended to deliver a range of capabilities, including intrusion detection, intrusion prevention, analytics, and information sharing. This program was developed to be one of the tools to aid federal agencies in mitigating information security threats.

[45] As part of the CyberStat Review, DHS conducted a Federal Incident Response Evaluation assessment in October 2017. The purpose of the assessment was, in part, to review the Bureau's incident management practices and provide recommendations that, if addressed, would strengthen the Bureau's cybersecurity efforts.

[46] According to OMB, high value assets are those assets (such as federal information systems, information, and data) for which an unauthorized access, use, disclosure, disruption, modification, or destruction could cause a significant impact.

[47] Although all of the recommendations from DHS are intended to assist the Bureau to improve its overall cybersecurity efforts, a recommendation may not explicitly indicate that there is a specific vulnerability. However, a recommendation may identify an area where the Bureau's cybersecurity capabilities could be strengthened.

[48] The National Institute of Standards and Technology defined penetration testing as security testing in which the evaluators mimic real-world attacks in an attempt to identify ways to circumvent the security features of an application, system, or network. Penetration testing often involves issuing real attacks on real systems and data, using the same tools and techniques used by actual attackers.

systems, and phishing assessments to gain access to sensitive PII. Of the 42 recommendations, 10 recommendations resulted from DHS's mandatory services for the Bureau (e.g., risk management and vulnerability assessments for specific high value assets). The remaining 32 recommendations resulted from DHS's voluntary services for the Bureau (e.g., Federal Incident Response Evaluation assessment). Due to the sensitive nature of the recommendations, we are not identifying the specific recommendations or specific findings associated with them in this statement.

In April 2019, we reported that the Bureau had not established a formal process for documenting, tracking, and completing corrective actions for all of the recommendations provided by DHS.[49] Accordingly, we recommended that the Bureau implement a formal process for tracking and executing appropriate corrective actions to remediate cybersecurity findings identified by DHS. As of late May 2019, the Bureau was working to address our recommendation.

Until the Bureau implements our recommendation, it faces an increased likelihood that findings identified by DHS will go uncorrected and may be exploited to cause harm to agency's 2020 Census IT systems and gain access to sensitive respondent data. Implementing a formal process would also help to ensure that DHS's efforts result in improvements to the Bureau's cybersecurity posture.

The Bureau Faces Several Other Cybersecurity Challenges in Implementing the 2020 Census

The Bureau faces other substantial cybersecurity challenges in addition to those previously discussed. More specifically, we previously reported that the extensive use of IT systems to support the 2020 Census redesign may help increase efficiency, but that this redesign introduces critical cybersecurity challenges.[50] These challenges include those related to the following:

---

[49] GAO-19-431T.
[50] GAO, *Information Technology: Better Management of Interdependencies between Programs Supporting 2020 Census Is Needed*, GAO-16-623 (Washington, D.C.: Aug. 9, 2016) and

- Phishing. We have previously reported that advanced persistent threats may be targeted against social media web sites used by the federal government. In addition, attackers may use social media to collect information and launch attacks against federal information systems through social engineering, such as phishing.[51] Phishing attacks could target respondents, as well as Bureau employees and contractors. The 2020 Census will be the first one in which respondents will be heavily encouraged to respond via the internet. This will likely increase the risk that cyber criminals will use phishing in an attempt to steal personal information. According to the Bureau, it plans to inform the public of the risks associated with phishing through its education and communication campaigns.
- Disinformation from social media. We previously reported that one of the Bureau's key innovations for the 2020 Census is the large-scale implementation of an internet self-response option. The Bureau is encouraging the public to use the internet self-response option through expanded use of social media. However, the public perception of the Bureau's ability to adequately safeguard the privacy and confidentiality of the 2020 Census internet self-responses could be influenced by disinformation spread through social media.

According to the Bureau, if a substantial segment of the public is not convinced that the Bureau can safeguard public response data against data breaches and unauthorized use, then response rates may be lower than projected, leading to an increase in cases for follow-up and subsequent cost increases. To help address this challenge, the Bureau stated that it plans to inform the public of the risks associated with disinformation from social media through its education and communication campaigns.

Information Technology: Uncertainty Remains about the Bureau's Readiness for a Key Decennial Census Test, GAO-17-221T (Washington, D.C.: Nov. 16, 2016).

[51] GAO, *Social Media: Federal Agencies Need Policies and Procedures for Managing and Protecting Information They Access and Disseminate,* GAO-11-605 (Washington, D.C.: June 28, 2011).

- Ensuring that individuals gain only limited and appropriate access to 2020 Census data. The Bureau plans to enable a public-facing website and Bureau-issued mobile devices to collect PII (e.g., name, address, and date of birth) from the nation's entire population— estimated to be over 300 million. In addition, the Bureau is planning to obtain and store administrative records containing PII from other government agencies to help augment information that enumerators did not collect.

The number of reported security incidents involving PII at federal agencies has increased dramatically in recent years. Because of these challenges, we have recommended, among other things, that federal agencies improve their response to information security incidents and data breaches involving PII, and consistently develop and implement privacy policies and procedures. Accordingly, it will be important for the Bureau to ensure that only respondents and Bureau officials are able to gain access to this information, and enumerators and other employees only have access to the information needed to perform their jobs.

- Ensuring adequate control in a cloud environment. The Bureau has decided to use cloud solutions as a key component of the 2020 Census IT infrastructure. We have previously reported that cloud computing has both positive and negative information security implications and, thus, federal agencies should develop service-level agreements with cloud providers.

These agreements should specify, among other things, the security performance requirements—including data reliability, preservation, privacy, and access rights—that the service provider is to meet.[52] Without these safeguards, computer systems and networks, as well as the critical operations and key infrastructures they support, may be lost; information—

---

[52] GAO, *Information Security: Agencies Need to Improve Cyber Incident Response Practices*, GAO-14-354 (Washington, D.C.: Apr. 30, 2014).

including sensitive personal information—may be compromised; and the agency's operations could be disrupted.

Commerce's Office of the Inspector General recently identified several challenges the Bureau may face using cloud-based systems to support the 2020 Census.[53] Specifically, in June 2019, the Office of the Inspector General identified, among other things, unimplemented security system features that left critical 2020 Census systems vulnerable during the 2018 End-to-End Test and a lack of fully implemented security practices to protect certain data hosted in the 2020 Census cloud environment. Officials from the Bureau agreed with all eight of the Office of Inspector General's recommendations regarding 2020 Census cloud-based systems and identified actions taken to address them.

- Ensuring contingency and incident response plans are in place to encompass all of the IT systems to be used to support the 2020 Census. Because of the brief time frame for collecting data during the 2020 Census, it is especially important that systems are available for respondents to ensure a high response rate. Contingency planning and incident response help ensure that, if normal operations are interrupted, network managers will be able to detect, mitigate, and recover from a service disruption while preserving access to vital information.

Implementing important security controls, including policies, procedures, and techniques for contingency planning and incident response, helps to ensure the confidentiality, integrity, and availability of information and systems, even during disruptions of service. Without contingency and incident response plans, system availability might be impacted and result in a lower response rate.

The Bureau's CIO has acknowledged these cybersecurity challenges and is working to address them, according to Bureau documentation. In

---

[53] U.S. Department of Commerce, Office of Inspector General, *The Census Bureau Must Correct Fundamental Cloud Security Deficiencies in Order to Better Safeguard the 2020 Decennial Census*, OIG-19-015-A (Washington, D.C.: June 19, 2019).

addition, we have ongoing work looking at many of these challenges, including the Bureau's plans to protect PII, use a cloud-based infrastructure, and recover from security incidents and other disasters.

Key Risk #4: The Bureau Will Need to Control Any Further Cost Growth and Develop Cost Estimates That Reflect Best Practices

Since 2015, the Bureau has made progress in improving its ability to develop a reliable cost estimate. We have reported on the reliability of the $12.3 billion life-cycle cost estimate released in October 2015 and the $15.6 billion revised cost estimate released in October 2017.[54] In 2016 we reported that the October 2015 version of the Bureau's life-cycle cost estimate for the 2020 Census was not reliable. Specifically, we found that the 2020 Census life-cycle cost estimate partially met two of the characteristics of a reliable cost estimate (comprehensive and accurate) and minimally met the other two (well-documented and credible). We recommended that the Bureau take specific steps to ensure its cost estimate meets the characteristics of a high-quality estimate. The Bureau agreed and has taken action to improve the reliability of the cost estimate.

In August 2018 we reported that while improvements had been made, the Bureau's October 2017 cost estimate for the 2020 Census did not fully reflect all the characteristics of a reliable estimate (See Figure 6).

In order for a cost estimate to be deemed reliable as described in GAO's Cost Estimating and Assessment Guide[55] and thus, to effectively inform 2020 Census annual budgetary figures, the cost estimate must meet or substantially meet the following four characteristics:

---

[54] GAO-16-628 and GAO, *2020 Census: Census Bureau Improved the Quality of Its Cost Estimation but Additional Steps Are Needed to Ensure Reliability*, GAO-18-635 (Washington, D.C.: Aug. 17, 2018).

[55] GAO, *GAO Cost Estimating and Assessment Guide: Best Practices for Developing and Managing Capital Program Costs (Supersedes GAO-07-1134SP)*, GAO-09-3SP (Washington, D.C.: Mar. 2, 2009).

Characteristic	2015 Assessment	2017 Assessment
Well-Documented	Minimally met	Partially met
Accurate	Partially met	Substantially met
Credible	Minimally met	Substantially met
Comprehensive	Partially met	Met

● Met   ◕ Substantially met   ◐ Partially met   ◔ Minimally met   ○ Not met

Source: GAO analysis of Census Bureau data. | GAO-19-588T.

Figure 6. Overview of the Census Bureau's 2015 and 2017 Cost Estimates Compared to Characteristics of a Reliable Cost Estimate.

- Well-Documented. Cost estimates are considered valid if they are well-documented to the point they can be easily repeated or updated and can be traced to original sources through auditing, according to best practices.
- Accurate. Accurate estimates are unbiased and contain few mathematical mistakes.
- Credible. Credible cost estimates must clearly identify limitations due to uncertainty or bias surrounding the data or assumptions, according to best practices.
- Comprehensive. To be comprehensive an estimate should have enough detail to ensure that cost elements are neither omitted nor double-counted, and all cost-influencing assumptions are detailed in the estimate's documentation, among other things, according to best practices.

The 2017 cost estimate only partially met the characteristic of being well-documented. In general, some documentation was missing, inconsistent, or difficult to understand. Specifically, we found that source data did not always support the information described in the basis of estimate document or could not be found in the files provided for two of the Bureau's largest field operations: Address Canvassing and Non-Response Follow-Up. We also found that some of the cost elements did not trace clearly to supporting spreadsheets and assumption documents.

Failure to document an estimate in enough detail makes it more difficult to replicate calculations, or to detect possible errors in the estimate; reduces transparency of the estimation process; and can undermine the ability to use the information to improve future cost estimates or even to reconcile the estimate with another independent cost estimate. The Bureau told us it would continue to make improvements to ensure the estimate is well-documented.

Increased Costs Are Driven by an Assumed Decrease in Self-Response Rates and Increases in Contingency Funds and IT Cost Categories

The 2017 life-cycle cost estimate includes much higher costs than those included in the 2015 estimate. The largest increases occurred in the Response, Managerial Contingency, and Census/Survey Engineering categories. For example, increased costs of $1.3 billion in the response category (costs related to collecting, maintaining, and processing survey response data) were in part due to reduced assumptions for self-response rates, leading to increases in the amount of data collected in the field, which is more costly to the Bureau.

Contingency allocations increased overall from $1.35 billion in 2015 to $2.6 billion in 2017, as the Bureau gained a greater understanding of risks facing the 2020 Census. Increases of $838 million in the Census/Survey Engineering category were due mainly to the cost of an IT contract for integrating decennial survey systems that was not included in the 2015 cost estimate.

Bureau officials attribute a decrease of $551 million in estimated costs for Program Management to changes in the categorization of costs associated with risks.

Specifically, in the 2017 version of the estimate, estimated costs related to program risks were allocated to their corresponding work breakdown structure (WBS) element. Figure 7 shows the change in cost by WBS category for 2015 and 2017.

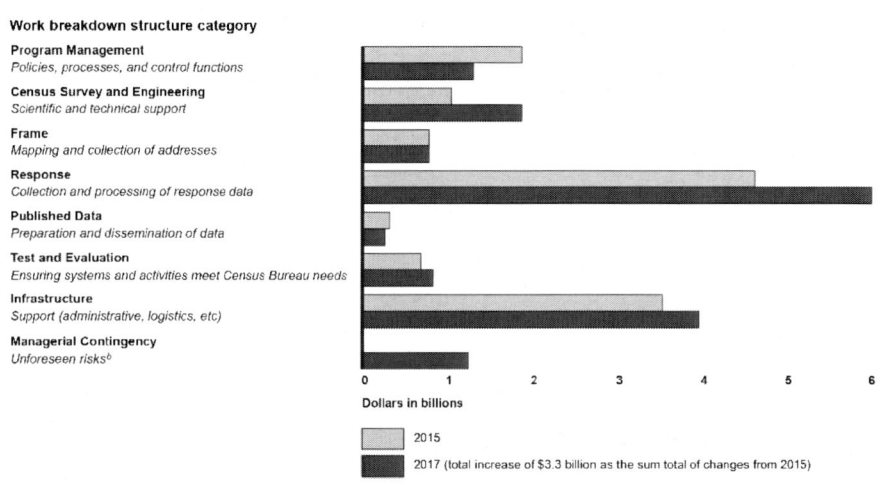

Source: GAO analysis of Census Bureau data. | GAO-19-588T.

[a]The historical life-cycle cost figures for prior decennials as well as the initial estimate for 2020 provided by Commerce in October 2017 differ slightly from those reported by the Bureau previously. According to Commerce documents, the more recently reported figures are "inflated to the current 2020 Census time frame (fiscal years 2012 to 2023)," rather than to constant 2020 dollars as the earlier figures had been. Specifically, since October 2017, Commerce and the Bureau have reported the October 2015 estimate for the 2020 Census as $12.3 billion; this is slightly different from the $12.5 billion the Bureau had initially reported.

[b]The 2015 cost estimate also included managerial contingency amounts totaling $829 million; however, these were not presented as a separate work breakdown structure category.

Figure 7. Change in 2020 Census Cost Estimate by Work Breakdown Structure Category, 2015 vs. 2017[a].

More generally, factors that contributed to cost fluctuations between the 2015 and 2017 cost estimates include:

- Changes in assumptions. Among other changes, a decrease in the assumed rate for self-response from 63.5 percent in 2015 to 60.5 percent in 2017 increased the cost of collecting responses from nonresponding housing units.
- Improved ability to anticipate and quantify risk. In general, contingency allocations designed to address the effects of potential risks increased overall from $1.3 billion in 2015 to $2.6 billion in 2017.
- An overall increase in IT costs. IT cost increases, totaling $1.59 billion, represented almost 50 percent of the total cost increase from 2015 to 2017.
- More defined contract requirements. Bureau documents described an overall improvement in the Bureau's ability to define and specify contract requirements. This resulted in updated estimates for several contracts, including for the Census Questionnaire Assistance contract.[56]

However, while the Bureau has been able to better quantify risk; in August 2018 we also reported that the Secretary of Commerce included a contingency amount of about $1.2 billion in the 2017 cost estimate to account for what the Bureau refers to as "unknown unknowns." According to Bureau documentation these include such risks as natural disasters or cyber-attacks. The Bureau provides a description of how the risk contingency for "unknown unknowns" is calculated; however, this description does not clearly link calculated amounts to the risks themselves. Thus, only $14.4 billion of the Bureau's $15.6 billion cost estimate has justification.

---

[56] This contract has two primary functions: to provide (1) questionnaire assistance by telephone and email for respondents by answering questions about the census in general and regarding specific items on the census form, and (2) an option for respondents to complete a census interview over the telephone.

According to Bureau officials, the cost estimate remains at $15.6 billion; however, they stated that they are managing the 2020 Census at a lower level of funding—$14.1 billion. In addition, they said that, at this time, they do not plan to request funding for the $1.2 billion contingency fund for unknown unknowns or $369 million in funding for selected discrete program risks for what-if scenarios, such as an increase in the wage rate or additional supervisors needed to manage field operations. Instead of requesting funding for these contingencies upfront the Bureau plans to work with OMB and Commerce to request additional funds, if the need arises.

According to Bureau officials they anticipate that the remaining $1.1 billion in contingency funding included in the $14.1 billion will be sufficient to carry out the 2020 Census. In June 2016 we recommended the Bureau improve control over how risk and uncertainty are accounted for. This prior recommendation remains valid given the life-cycle cost estimate still includes the $1.2 billion unjustified contingency fund for "unknown unknowns".

Moreover, given the cost growth between 2015 and 2017 it will be important for the Bureau to monitor cost in real-time, as well as, document, explain and review variances between planned and actual cost. In August 2018 we reported that the Bureau had not been tracking variances between estimated life-cycle costs and actual expenses. Tools to track variance enable management to measure progress against planned outcomes and will help inform the 2030 Census cost estimate. Bureau officials stated that they already have systems in place that can be adapted for tracking estimated and actual costs. We will continue to monitor the status of the tracking system.

According to Bureau officials, the Bureau planned to release an updated version of the 2020 Census life-cycle estimate in the spring of 2019; however, they had not done so as of June 28, 2019. To ensure that future updates to the life-cycle cost estimate reflect best practices, it will be important for the Bureau to implement our recommendation related to the cost estimate.

CONTINUED MANAGEMENT ATTENTION NEEDED TO KEEP PREPARATIONS ON TRACK AND HELP ENSURE A COST-EFFECTIVE ENUMERATION

2020 Challenges Are Symptomatic of Deeper Long-Term Organizational Issues

The difficulties facing the Bureau's preparation for the decennial census in such areas as planning and testing; managing and overseeing IT programs, systems, and contractors supporting the enumeration; developing reliable cost estimates; prioritizing decisions; managing schedules; and other challenges, are symptomatic of deeper organizational issues.

Following the 2010 Census, a key lesson learned for 2020 that we identified was ensuring that the Bureau's organizational culture and structure, as well as its approach to strategic planning, human capital management, internal collaboration, knowledge sharing, capital decision-making, risk and change management, and other internal functions are aligned toward delivering more cost-effective outcomes.[57]

The Bureau has made improvements over the last decade, and continued progress will depend in part on sustaining efforts to strengthen risk management activities, enhancing systems testing, bringing in experienced personnel to key positions, implementing our recommendations, and meeting regularly with officials from its parent agency, Commerce.

Going forward, we have reported that the key elements needed to make progress in high-risk areas are top-level attention by the administration and agency officials to (1) leadership commitment, (2) ensuring capacity, (3) developing a corrective action plan, (4) regular monitoring, and (5) demonstrated progress. Although important steps have been taken in at

---

[57] GAO, *2010 Census: Preliminary Lessons Learned Highlight the Need for Fundamental Reforms*, GAO-11-496T (Washington, D.C.: Apr. 6, 2011).

least some of these areas, overall, far more work is needed.[58] We discuss three of five areas below.

The Secretary of Commerce has successfully demonstrated leadership commitment. For example, the Bureau and Commerce have strengthened this area with executive-level oversight of the 2020 Census by holding regular meetings on the status of IT systems and other risk areas. In addition, in 2017 Commerce designated a team to assist senior Bureau management with cost estimation challenges. Moreover, on January 2, 2019, a new Director of the Census Bureau took office, a position that had been vacant since June 2017.

With regard to capacity, the Bureau has improved the cost estimation process of the decennial when it established guidance including:

- roles and responsibilities for oversight and approval of cost estimation processes,
- procedures requiring a detailed description of the steps taken to produce a high-quality cost estimate, and
- a process for updating the cost estimate and associated documents over the life of a project.

However, the Bureau continues to experience skills gaps in the government program management office overseeing the $886 million contract for integrating the IT systems needed to conduct the 2020 Census. Specifically, as of June 2019, 14 of 44 positions in this office were vacant.

For the monitoring element, we found to track performance of decennial census operations, the Bureau relied on reports to track progress against pre-set goals for a test conducted in 2018. According to the Bureau, these same reports will be used in 2020 to track progress. However, the Bureau's schedule for developing IT systems during the 2018 End-to-End test experienced delays that compressed the time available for system testing, integration testing, and security assessments. These schedule delays contributed to systems experiencing problems after deployment, as

---

[58] GAO-17-317.

well as cybersecurity challenges. In the months ahead, we will continue to monitor the Bureau's progress in addressing each of the five elements essential for reducing the risk to a cost-effective enumeration.

FURTHER ACTIONS NEEDED ON OUR RECOMMENDATIONS

Over the past several years we have issued numerous reports that underscored the fact that, if the Bureau was to successfully meet its cost savings goal for the 2020 Census, the agency needed to take significant actions to improve its research, testing, planning, scheduling, cost estimation, system development, and IT security practices. As of June 2019, we have made 106 recommendations related to the 2020 Census. The Bureau has implemented 74 of these recommendations, 31 remain open, and one recommendation was closed as not implemented.

Of the 31 open recommendations, 9 were directed at improving the implementation of the innovations for the 2020 Census. Commerce generally agreed with our recommendations and is taking steps to implement them. Moreover, in April 2019 we wrote to the Secretary of Commerce, providing a list of the 12 open 2020-Census-related recommendations that we designated as "priority."[59] Priority recommendations are those recommendations that we believe warrant priority attention from heads of key departments and agencies.

We believe that attention to these recommendations is essential for a cost-effective enumeration. The recommendations included implementing reliable cost estimation and scheduling practices in order to establish better control over program costs, as well as taking steps to better position the Bureau to develop an internet response option for the 2020 Census.

In addition to our recommendations, to better position the Bureau for a more cost-effective enumeration, on March 18, 2019, we met with OMB,

---

[59] The 12 priority recommendations originated in reports we issued from November 2009 to December 2018.

Commerce, and Bureau officials to discuss the Bureau's progress in reducing the risks facing the census. We also meet regularly with Bureau officials and managers to discuss the progress and status of open recommendations related to the 2020 Census, which has resulted in Bureau actions in recent months leading to closure of some recommendations.

We are encouraged by this commitment by Commerce and the Bureau in addressing our recommendations. Implementing our recommendations in a complete and timely manner is important because it could improve the management of the 2020 Census and help to mitigate continued risks.

In conclusion, while the Bureau has made progress in revamping its approach to the census, it faces considerable challenges and uncertainties in implementing key cost-saving innovations and ensuring they function under operational conditions; managing the development and testing of its IT systems; ensuring the cybersecurity of its systems and data; and developing a quality cost estimate for the 2020 Census and preventing further cost increases. For these reasons, the 2020 Census is a GAO high-risk area.

Going forward, continued management attention and oversight will be vital for ensuring that risks are managed, preparations stay on track, and the Bureau is held accountable for implementing the enumeration, as planned. Without timely and appropriate actions, the challenges previously discussed could adversely affect the cost, accuracy, schedule, and security of the enumeration. We will continue to assess the Bureau's efforts and look forward to keeping Congress informed of the Bureau's progress.

Chairman Johnson, Ranking Member Peters, and Members of the Committee, this completes our prepared statement. We would be pleased to respond to any questions that you may have.

In: Actions Needed ...
Editor: Bryant Schneider

ISBN: 978-1-53616-716-0
© 2019 Nova Science Publishers, Inc.

Chapter 3

2020 CENSUS: PROGRESS REPORT ON THE CENSUS BUREAU'S EFFORTS TO CONTAIN ENUMERATION COSTS[*]

Robert Goldenkoff and Carol R. Cha

WHY GAO DID THIS STUDY

At $13 billion, 2010's headcount was the costliest in U.S. history. Thus, over the next few years, the fundamental challenge facing Bureau leadership will be designing and implementing a census that controls the cost of the enumeration while maintaining its accuracy.

This chapter focuses on progress the Bureau is making in three areas key to a more cost-effective enumeration: (1) transforming the Bureau into a higher-performing organization; (2) improving the cost-effectiveness of

---

[*] This is an edited, reformatted and augmented version of a report, GAO-13-857T, issued by the U.S. Government Accountability Office, September 11, 2013. This report was originally presented as testimony before the House Committee on Oversight and Government Reform, Subcommittee on Federal Workforce, U.S. Postal Service, and the Census.

census-taking operations; and (3) strengthening IT management and security practices. This chapter is based on completed work that included an analysis of Bureau documents, interviews with Bureau officials, and field observations of census operations in urban and rural locations across the country.

WHAT GAO RECOMMENDS

GAO is not making new recommendations in this testimony but reports on the status of past recommendations that the Bureau strengthen its IT management, develop policies and procedures for its cost estimates, and integrate its 2020 Census planning. The Bureau generally agreed with GAO's findings and recommendations and is taking steps to implement them.

WHAT GAO FOUND

In preparing for the 2020 Census, the U.S. Census Bureau (Bureau) has launched several initiatives aimed at organizational transformation, some of which show particular promise. For example, the Bureau is attempting to develop Bureau-wide, or "enterprise," standards, guidance, or tools in areas such as risk management and information technology (IT) investment management to reduce duplicative efforts across the Bureau. Although the Bureau has made progress in these and other areas, if the Bureau is to transform itself to better control costs and deliver an accurate national headcount in 2020, several areas will require continued oversight: cost estimation, integrated long-term planning, and stakeholder involvement. For example, while the Bureau has made progress with long-term planning by implementing some elements of GAO's recommendation that it develop a road map for 2020 planning, it still needs to pull together remaining planning elements, such as milestones for decisions and estimates of cost, into its roadmap.

The Bureau is researching several key operational initiatives that may yield significant cost savings. However, while these initiatives have the potential to reduce costs, the Bureau will be employing them in ways that are new for 2020 and thus entail some operational risk. Key among these are using the Internet as a self-response option, targeting only certain addresses for field verification as the Bureau builds its national list of addresses, and replacing enumerator-collected data with administrative records under certain circumstances. Bureau tests conducted in 2011 showed that adding an Internet response option to the census could increase its overall response rate, which could save money, since Bureau field staff would need to visit fewer households, which is its largest and most costly census field operation. In addition, the Bureau has estimated that it could save up to $2 billion if it uses administrative records in 2020 to reduce the need for related costly and labor-intensive door-to-door visits by Bureau employees.

Additionally, the Bureau is exploring technology options for census operations that collectively represent a dramatic leap from 2010. These options include the possible use of a "bring your own device" model to enable enumerators to use their own mobile devices for field data collection. Given the role of information technology in conducting the census, while controlling cost and protecting privacy, it is essential that the Bureau strengthen its ability to manage these investments, as well as its practices for securing the information it collects and disseminates. The Bureau faces several long-standing IT challenges that, if effectively addressed, will significantly enhance its ability to acquire these solutions within cost, schedule, and performance targets. For example, effective workforce planning is essential to ensuring organizations have the proper skills, abilities, and capacity for effective IT management; however, the Bureau has not yet finalized its IT workforce plans. Additionally, in January 2013, GAO reported that controls over access to the Bureau's IT systems contained deficiencies. Without adequate system access controls, the Bureau cannot be sure that its information and systems are protected from intrusion.

Chairman Farenthold, Ranking Member Lynch, and Members of the Subcommittee:

We are pleased to participate in today's hearing to discuss the U.S. Census Bureau's (Bureau) preparations for the next enumeration. Although Census Day 2020 is still more than 6 years away, research and testing activities for the decennial have been progressing for some time, and the Bureau will be making key design decisions in 2014 and 2015. Our reviews of the 1990, 2000, and 2010 enumerations underscore the importance of early planning and strong and continuing congressional oversight to reduce the costs and risks of the national headcount as well as to keep the entire enterprise on track.

At $13 billion, 2010's headcount was the costliest in U.S. history. Thus, over the next few years, the fundamental challenge facing Bureau leadership will be designing and implementing a census that simultaneously controls the cost of the enumeration while maintaining its accuracy.

The basic design of the enumeration—mail out and mail back of the census questionnaire with in-person follow-up for nonrespondents—has been in use since 1970. A key lesson learned from 2010 and earlier enumerations is that this design is no longer capable of cost-effectively counting a population that is growing steadily larger, more diverse, increasingly difficult to find, and reluctant to participate in the census. The Bureau is well aware that reforms are needed, and plans to significantly change the methods and technologies it uses to enumerate the population. However, the Bureau has never before employed many of these methods at the scale being considered for 2020, if at all, which adds a large degree of risk. Moreover, the Bureau's past efforts to implement new approaches and systems have not always gone well. As one example, during the 2010 Census the Bureau planned to use handheld mobile devices to support field data collection for the census, including following up with nonrespondents. However, due to significant problems identified during testing of the devices, cost overruns, and schedule slippages, the Bureau decided not to use the handheld devices for non-response follow-up and reverted to paper-based processing, which increased the cost of the 2010 Census by up to $3 billion and significantly added to its risk as it had to switch its operations to paper-based operations as its backup.

As the Bureau launched its preparations for 2020 earlier this decade, we noted that controlling census costs while maintaining accuracy hinged on the Bureau addressing challenges in three key areas: (1) transforming the Bureau into a high-performing organization; (2) improving the cost-effectiveness of census-taking operations; and (3) strengthening information technology (IT) management and security practices. With this as backdrop, our remarks this morning will focus on the Bureau's plans for 2020, paying particular attention to the status of cost-containment initiatives within each of these three areas. In particular we will discuss where the Bureau has made progress, and management challenges and open questions that the Bureau will need to resolve going forward.

In summary, we found that the Bureau is progressing along a number of fronts to secure a more cost-effective enumeration. For example, the Bureau's organizational transformation efforts, which includes efforts to improve its workforce in order to help the Bureau become more results oriented.

At the same time, innovative enumeration methods such as the use of administrative records to assist with enumerating people, use of the Internet to collect data, and targeted address canvassing might help to control costs, but a number of operational uncertainties remain, such as ensuring privacy and information security with some of the new approaches. Likewise, the Bureau's ability to effectively and efficiently acquire the technological solutions supporting 2020 will be largely dependent on having established, mature IT management controls, an area of long-standing concern to us.

The information in our testimony is based on our previous reports on the 2010 Census, as well as the Bureau's planning efforts for 2020.[1] For this

---

[1] See related GAO products at the end of this statement.

work, among other things we analyzed key documents such as budgets, plans, procedures, and guidance for selected activities; and interviewed cognizant Bureau officials at headquarters and local census offices. In addition, for the work on the 2010 Census, we made on-site observations of key enumeration activities across the country including both urban and less populated areas. To obtain information on various management and organizational reforms that could help the Bureau become more accountable and results oriented, we reviewed our prior work on government-wide reexamination, as well as leading practices and attributes in the areas of IT management, organizational performance, collaboration, stewardship, and human capital.[2] More detail on our scope and methodology is provided in each published report that this testimony is based on.

We provided the Bureau with a summary of the information included in this statement, and Bureau officials provided technical comments, which we included as appropriate. We conducted the work that this testimony is based on in accordance with generally accepted government auditing standards. Those standards require that we plan and perform the audit to obtain sufficient, appropriate evidence to provide a reasonable basis for our findings and conclusions based on our audit objectives. We believe that the evidence obtained provides a reasonable basis for our findings and conclusions based on our audit objectives.

---

[2] See for example: GAO, *Results-Oriented Government: Practices That Can Help Enhance and Sustain Collaboration among Federal Agencies*, GAO-06-15 (Washington, D.C.: Oct. 21, 2005); *21st Century Challenges: Reexamining the Base of the Federal Government*, GAO-05-325SP (Washington, D.C.: February 2005); *Information Technology Investment Management: A Framework for Assessing and Improving Process Maturity*, GAO-04-394G (Washington, D.C.: March 2004); *Comptroller General's Forum, High-Performing Organizations: Metrics, Means, and Mechanisms for Achieving High Performance in the 21st Century Public Management Environment*, GAO-04-343SP (Washington, D.C.: Feb. 13, 2004; and *Human Capital: Key Principles for Effective Strategic Workforce Planning*, GAO-04-39 (Washington, D.C.: Dec. 11, 2003).

BACKGROUND

The decennial census is mandated by the U.S. Constitution and provides data that are vital to the nation. This information is used to apportion the seats of the U.S. House of Representatives; realign the boundaries of the legislative districts of each state; allocate billions of dollars in federal financial assistance; and provide social, demographic, and economic profiles of the nation's people to guide policy decisions at each level of government.

Although the complexity, cost, and importance of the census necessitate robust planning, recent enumerations were not planned well. Our prior work has found shortcomings with managing, planning, and implementing IT solutions in the 2000 and 2010 enumerations that led to acquisition problems, cost overruns, and other issues. As a result, we placed both enumerations on our list of high-risk programs.[3] For example, leading up to the 2010 Census, we found that the lack of skilled cost estimators for the 2010 Census led to unreliable life-cycle cost estimates, and some key operations were not tested under census-like conditions.

As shown in figure 1, the cost of enumerating each housing unit has escalated from around $16 in 1970 to around $98 in 2010, in constant 2010 dollars (an increase of over 500 percent). At the same time, the mail response rate—a key indicator of a cost-effective enumeration—has declined from 78 percent in 1970 to 63 percent in 2010. In many ways, the Bureau has had to invest substantially more resources each decade just to try and match the results of prior enumerations.

Beginning in 1990, we reported that rising costs, difficulties in securing public participation, and other long-standing challenges required a revised census methodology—a view that was shared by other stakeholders.[4] Since then, we and other organizations—including the

---

[3] GAO, *Information Technology: Significant Problems of Critical Automation Program Contribute to Risks Facing 2010 Census*, GAO-08-550T (Washington, D.C.: Mar. 5, 2008) and *High-Risk Series: Quick Reference Guide*, GAO/HR-97-2 (Washington, D.C.: February 1997).

[4] See for example, GAO, *2000 Census: Progress Made on Design, but Risks Remain*, GAO/GGD-97-142 (Washington, D.C.: July 14, 1997), and *Decennial Census: Preliminary*

Bureau itself— have stated that fundamental changes to the design, implementation, and management of the census must be made in order to address operational and organizational challenges.[5] In response, the Bureau has stated that containing costs and maintaining quality will require bold innovations in the planning and design of the 2020 Census. The Bureau has also stated its goal is to conduct the 2020 Census at a lower cost per housing unit than the approximately $98 per housing unit cost of the 2010 Census (in constant 2010 dollars) while still maintaining high quality.

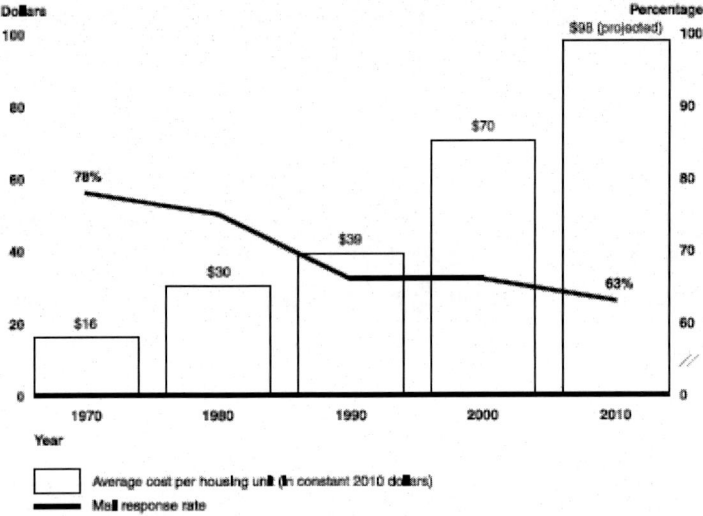

Source: GAO analysis of Census Bureau Data.

Note: In the 2010 Census the Bureau used only a short-form questionnaire. For this statement, we use the 1990 and 2000 Census short-form mail response rate when comparing 1990, 2000, and 2010 mail-back response rates. Census short-form mail response rates are unavailable for 1970 and 1980, so we use the overall response rate.

Figure 1. The Average Cost of Counting Each Housing Unit (in Constant 2010 Dollars) Has Escalated Each Decade while Mail Response Rates Have Declined.

1990 Lessons Learned Indicate Need to Rethink Census Approach, GAO/T-GGD-90-18 (Washington, D.C.: Aug. 8, 1990).

[5] GAO, *2020 Census: Sustaining Current Reform Efforts Will Be Key to a More Cost-Effective Enumeration*, GAO-12-905T (Washington, D.C.: July 18, 2012).

THE BUREAU'S PLANS FOR CONTROLLING ENUMERATION COSTS SHOW PROMISE, BUT KEY CHALLENGES NEED TO BE ADDRESSED

Transforming the Bureau into a High-Performing Organization

The Bureau's experience with the 2010 and prior enumerations has shown that lack of proper planning and not following leading practices in key management areas can increase the costs and risks of later downstream operations. For example, in a self-assessment in October 2008, the Bureau found that its organizational structure made overseeing a large program difficult and hampered accountability, succession planning, and staff development. Moreover, leading up to the 2010 Census, we reported that internal organizational, planning, funding, and human capital challenges jeopardized the Bureau's overall readiness.

In preparing for 2020, the Bureau has launched several initiatives aimed at organizational transformation, some of which show particular promise if successfully implemented.

- Organizational restructuring. The Bureau's organizational transformation took a significant step forward in July 2011 when it created a 2020 Census Directorate that included the office responsible for the American Community Survey, supporting the Bureau's objective to rely on that nation-wide survey as a "test bed" for cost saving innovations for the 2020 Census. The Bureau is undertaking an organizational transformation of its entire decennial directorate in order to improve collaboration and communication across its divisions, improve operational efficiencies, and instill a culture that, according to the Bureau, encourages risk-taking and innovation without fear of reprisal. The Bureau believes such change is necessary so that it can more effectively control costs and enumerate the population for 2020.

- Enterprise solutions. The Bureau is attempting to develop Bureau-wide, or "enterprise," standards, guidance, or tools in areas such as risk management, project management, systems engineering, and IT investment management in order to reduce duplicative efforts across the Bureau.
- Better workforce planning. As the Bureau reexamines how it will plan the 2020 Census, it is also reviewing the employee skills and competencies needed to make that happen, in part by a formal analysis comparing its needs to its in-house capabilities.

The Bureau has made progress in these areas and others. However, they will require continued oversight if the Bureau is to transform itself to better control costs and deliver an accurate national headcount in 2020.

Cost Estimation

Our prior work has highlighted the need for the Bureau to develop more accurate and rigorous cost estimates for census operations.[6] The Bureau uses the life-cycle cost estimate as the starting point for the annual budget formulation process and, according to our Cost Estimating and Assessment Guide, a reliable cost-estimating process is necessary to ensure that cost estimates—particularly for large, complex projects like the 2020 Census—are comprehensive, well documented, accurate, and credible.[7] In January 2012, among other actions, we recommended that the Bureau finalize guidance, policies, and procedures for cost estimation in accordance with best practices prior to developing the initial 2020 life-cycle cost estimate.[8]

In response to our recommendation, the Bureau has created a cost estimation team reporting to the Director. The team intended, among

---

[6] GAO, *2010 Census: Census Bureau Should Take Acton to Improve the Credibility and Accuracy of Its Cost Estimate for the Decennial Census*, GAO-08-554 (Washington, D.C.: June 16, 2008).

[7] GAO, *GAO Cost Estimating and Assessment Guide: Best Practices for Developing and Managing Capital Program Costs*, GAO-09-3SP (Washington, D.C.: March 2009).

[8] GAO, *Decennial Census: Additional Actions Could Improve the Census Bureau's Ability to Control Costs for the 2020 Census*, GAO-12-80 (Washington, D.C.: Jan. 24, 2012).

other things, to standardize guidance and training in cost estimation throughout the Bureau. The Bureau recently took the important step of hiring an individual to lead that group. However, until the Bureau finalizes its cost-estimating policies, procedures, and guidance, as we recommended, it runs the risks of developing unreliable cost estimates for 2020.

Integrated Long-Term Planning

The Bureau's progress thus far with early planning is noteworthy given its long-standing challenges in this area. In December 2010,[9] we recommended that the Bureau develop a roadmap for 2020 that integrates performance, budget, methodological, schedule, and other information that would be updated as needed and posted on the Bureau's website and other social media outlets. We also recommended that the Bureau develop a mechanism that allows for and harnesses input from census stakeholders and individuals. The Bureau agreed with our recommendations and brought together some of these elements in an annual fiscal year update of its "business plan," which it issued to Congress in concert with its budget submissions for each of the past 2 years. However, as the approach for 2020 takes shape, the Bureau needs to fully implement our recommendation to pull together remaining planning elements, such as milestones for decisions and estimates of cost into its tactical plan or roadmap.

In addition, we recommended in November 2009 that the Bureau improve its use of a master activity schedule for 2020 to include levels of resources and take other steps that would support systematic analyses of the risk to the schedule.[10] The Department of Commerce did not comment on that recommendation in its response to that report, but the Bureau has since developed an integrated schedule covering its early research and

---

[9] GAO, *2010 Census: Data Collection Operations Were Generally Completed as Planned, but Long-standing Challenges Suggest Need for Fundamental Reforms*, GAO-11-193 (Washington, D.C.: Dec. 14, 2010).

[10] See GAO, *2010 Census: Census Bureau Has Made Progress on Schedule and Operational Control Tools, but Needs to Prioritize Remaining System Requirements*, GAO-10-59 (Washington, D.C.: Nov. 13, 2009).

testing activity that we are reviewing as part of ongoing work. Implementing additional steps such as those we have recommended will help ensure the Bureau's reform initiatives stay on track, do not lose momentum, and coalesce into a viable path toward a more cost-effective 2020 Census.

Stakeholder Involvement

Ensuring active stakeholder involvement and buy-in is critical to high-performing organizations. For example, over the past decade we have reported on the importance of congressional outreach to secure early agreement between the Bureau and Congress on the Bureau's fundamental approach for its next decennial.[11]

In response to these reports and recommendations that we made, the Bureau has taken several steps forward. For example, in July 2012, the Bureau issued a plan for 2020 Census communications and stakeholder engagement, describing roles and responsibilities, among other elements. In December 2012, the Bureau began quarterly reviews intended to provide internal and external census program stakeholders, including congressional staff, officials from the Office of Management and Budget, and the Department of Commerce and its Office of Inspector General, with a broad and timely status of planning and development projects thereby facilitating strategic guidance and information sharing.[12] These are important strides by the Bureau to ensure its research and planning are transparent. However, the challenge remains for the Bureau to identify tradeoffs among cost, quality, privacy, and security that may arise in the Bureau's proposed approaches, and raise these tradeoffs with stakeholders.

Improving the Cost-Effectiveness of Census- Taking Operations

The Bureau's current research and testing phase represents a critical stage in preparing for a cost-effective 2020 Census. Bureau management

---

[11] See GAO, *2020 Census: Additional Steps Are Needed to Build on Early Planning*, GAO-12-626 (Washington, D.C.: May 17, 2012), and *2010 Census: Cost and Design Issues Need to be Addressed Soon*, GAO-04-37 (Washington, D.C.: Jan. 15, 2004).

[12] We also attend these reviews as observers at the Bureau's invitation.

will use the results of ongoing research and testing to shape the next decennial census as it determines what new operations will be a part of the 2020 Census design, which operations need to be revised, and how to mitigate remaining risks.

The Bureau may be able to use its research initiatives during the next couple of years to attain significant cost savings. Key among these are three new operational changes being considered—using the Internet as a self-response option, targeting only certain addresses for field verification as the Bureau builds its national list of addresses, and replacing enumerator-collected data with administrative records under certain circumstances. All three initiatives have the potential to reduce costs. However, the Bureau will be employing them in ways that are new for 2020, and they thus entail some operational risk. Going forward, the Bureau needs to ensure they will (1) produce needed cost savings, (2) function in concert with other census operations, and (3) work at the scale needed for the national headcount.

Using the Internet to Collect Responses

Tests conducted by the Bureau in 2011 showed that adding an Internet response option could increase the overall response rate for the census. The 2011 test results, coupled with the increased prevalence and accessibility of the Internet, led Bureau officials to commit to providing an Internet response option for the 2020 Census. If this option can help achieve an overall increase in the response rate, it can save money, since Bureau field staff would need to visit fewer households during nonresponse follow-up (NRFU), which is the largest and most costly census field operation.[13] Furthermore, testing has shown that the cost of an Internet survey is low compared to a mail survey, which incurs printing and postage costs. Moreover, web survey responses are generally available more quickly and

---

[13] During NRFU the Bureau sends enumerators to collect data from households that did not mail back their census forms. NRFU procedures instruct enumerators to make up to six attempts to contact a household. The 2010 Census NRFU operation cost $1.6 billion.

are of better quality than responses from a mail survey because there is no lag time, as the responses are captured in real time, and there are reminders to prompt the respondent if a question is unanswered. Quicker and more complete responses can also help reduce the amount of time and money spent on following up on late or incomplete census forms.

Targeting Address Canvassing

In the 2010 and earlier censuses, the Bureau mounted a full address canvassing operation, where field staff travelled virtually every road in the country to update the Master Address File (MAF) and the associated mapping database called TIGER (Topologically Integrated Geographic Encoding and Referencing). This labor-intensive effort was one of the more expensive components of the 2010 Census. It required 140,000 temporary workers to verify 145 million addresses (by going door-to-door) at a cost of $444 million, or 3 percent of the $13 billion total cost of the 2010 Census. For the 2020 Census the Bureau would like to reduce workload and cost by targeting the address canvassing operation to areas most in need of updating.

Administrative Records

Administrative records are a growing source of information on individuals and households. The Bureau has estimated that it could save up to $2 billion if it uses administrative records to reduce the need in 2020 for certain costly and labor-intensive door-to-door visits by Bureau employees, such as collecting data in person from nonrespondents, supporting quality control, or helping to evaluate the quality of the census.[14] For purposes of the decennial census, the Bureau is considering administrative records from government agencies, including tax data and Medicare records,[15] as

---

[14] The amount of and quality of administrative records the Bureau is able to collect will affect the amount of cost savings it is able to realize.

[15] The Bureau's access to and use of administrative records is governed by agency-specific statutes. For example, the Bureau has access to tax data under 26 U.S.C. § 6103(j)(1) "for the

well as commercial sources to identify persons associated with a particular household address. During the 2010 Census, the Bureau made limited use of administrative records. For example, the Bureau used U.S. Postal Service files to update its address list, and it used federal agency records (such as those from the Department of Defense) to count military and federal civilian employees stationed outside of the United States.

Depending on the results of ongoing research, Bureau officials plan to build a composite of quality administrative records from various sources (i.e., federal agencies, state and local governments, and commercial sources) that it can use to reduce or replace costly field work. Successful use of such a database presents challenges the Bureau will need to address. For example, as we reported in 2012, while the Bureau has access to some federally collected data, it does not have access to all of the federally collected administrative data that could potentially help it reduce the cost of the 2020 Census.[16] Further increasing the Bureau's access to records may involve negotiations with states or other federal agencies, potential statutory changes, and discussions of personal privacy protections, and most likely it would be a time-consuming process. In addition, the use of administrative records may present difficult decisions about tradeoffs between cost and quality, which the Bureau is actively researching to inform.

Strengthening IT Management and Security Practices

Additionally, the Bureau is exploring technology options for census operations that collectively represent a dramatic leap from 2010. These options include the possible use of a "bring your own device" model to enable enumerators to use their own mobile devices for field data collection. Given the role of information technology in conducting the census, while controlling costs and protecting privacy, it is essential that the Bureau strengthen its ability to manage these investments, as well as its

---

purpose of, but only to the extent necessary in, the structuring of censuses ... and conducting related statistical activities."

[16] GAO, *2020 Census: Initial Research Milestones Generally Met but Plans Needed to Mitigate Highest Risks*, GAO-13-53 (Washington, D.C.: Nov. 7, 2012).

practices for securing the information it collects and disseminates. The following represent long-standing IT challenges that, if effectively addressed, will significantly enhance the Bureau's ability to acquire these solutions within cost, schedule, and performance targets.

IT Governance

The Bureau lacks a sufficiently mature IT governance process to ensure that its investments are properly controlled and monitored. Implementing a governance framework and system development methodology are challenging tasks that can be aided by having robust implementation plans. Such a plan is instrumental in helping agencies coordinate and guide improvement efforts. In September 2012,[17] we reported that while the Bureau developed the Enterprise Investment Management Plan, which was to be applied to all investments, the plan was still a draft document and had key gaps. Specifically, the plan did not contain guidelines for the membership of investment review boards or the frequency of board meetings, and it omitted cost and schedule performance thresholds for escalating issues to higher-level boards. Accordingly, we made recommendations to address these weaknesses. The Bureau agreed, and in response to our recommendations, in June 2013, program officials provided us with an updated plan, which was finalized on September 28, 2012. However, while the plan now states that investment review boards should meet at least monthly, the plan does not specify thresholds for escalating cost, risk, or impact issues. The Bureau needs to take action in this key area as we previously recommended to ensure that its senior executives have adequate insight into project health to make timely decisions.

Requirements Management

Proper requirements management remains a long-standing challenge for the Bureau. The Software Engineering Institute states that a disciplined process for developing and managing requirements can help reduce the

---

[17] GAO, *Information Technology: Census Bureau Needs to Implement Key Management Practices*, GAO-12-915 (Washington, D.C.: Sept. 18, 2012).

risks of developing or acquiring a system. Unfortunately, the Bureau has had difficulties with this in the past, as illustrated by the problems it had in managing requirements during the 2010 census, which were largely responsible for the Bureau's abandonment of its handheld enumeration devices and increased the cost of the census by up to $3 billion. In September 2012,[18] we reported that the IT and 2020 Census directorates had independently drafted new requirements, instead of developing a Bureau-wide requirements management plan, despite our prior recommendation. To address the Bureau's recurring weaknesses in requirements management, we therefore recommended that it establish and implement a consistent requirements development and management process across the Bureau. Bureau officials agreed with the recommendation and in response, in August 2013, program officials stated that they began using a new life-cycle management tool to manage requirements Bureau-wide.

While this is a good start, it remains to be seen whether the Bureau will fully implement the new tool and institutionalize the requirements management process. Until the Bureau fully implements our recommendation to establish a consistent requirements development and management process across the Bureau that has clear guidance for developing requirements at the strategic mission, business, and project levels and is integrated with its new system development methodology, it will not have assurance that the IT systems delivered for 2020 will actually meet user needs.

IT Workforce Planning

As discussed earlier in this statement, effective workforce planning is essential to ensure organizations have the proper skills, abilities, and capacity for effective management. The Bureau has not yet finalized its IT workforce plans. In 2012, we reported that the Bureau had taken limited steps to develop IT human capital practices, such as inventorying critical competencies among its IT staff.[19] Yet many key steps remained to be

---

[18] GAO-12-915.
[19] GAO-12-915.

implemented. In particular, the Bureau had not developed a Bureau-wide IT workforce plan, identified gaps in mission-critical IT occupations, skills, and competencies, or developed strategies to address gaps. Accordingly, we recommended that the Bureau establish a repeatable process for performing IT skills assessments and gap analyses that can be implemented in a timely manner. The Bureau agreed with the recommendation, and in response, in June 2013, Bureau officials stated that they plan to complete a skills and needs assessment for the Bureau's IT workforce by the end of this month. Officials also reported that they have a workforce planning team that has developed a strategic workforce planning process and implementation plan. While the Bureau has taken certain steps to improve its IT workforce planning processes, going forward it will be important for it to fully establish a repeatable process for performing skills assessments and gap analyses, as we recommended, that can be implemented in a timely manner and better enable managers to address any skills gaps in preparation for the 2020 Census.

IT System Security

Critical to the Bureau's ability to perform its data collection and analysis duties are its information systems and the protection of the information they contain. A data breach could result in the public's loss of confidence in the Bureau, thus affecting its ability to collect census data. Access controls are designed and implemented to ensure the reliability of an agency's computerized information.[20] Access controls that are intended to prevent, limit, and detect unauthorized access to computing resources, programs, information, and facilities, are referred to as logical and physical access controls. Inadequate design or implementation of access controls increases the risk of unauthorized disclosure, modification, and destruction of sensitive information and disruption of service.

---

[20] Access controls include those related to (1) protection of system boundaries, (2) identification and authentication, (3) authorization, (4) cryptography, (5) audit and monitoring, and (6) physical security.

In January 2013, we reported that the Bureau's IT systems' access controls contained certain deficiencies.[21] For example, the Bureau did not adequately control connectivity to key network devices and servers, identify and authenticate users, or limit user access rights and permissions to only those necessary to perform official duties. An underlying reason for those weaknesses was that the Bureau had not fully implemented a comprehensive information security program to ensure that controls were effectively established and maintained. Accordingly, we recommended that the Bureau take several actions, such as clearly documenting its assessment of common controls for information systems before granting an authorization to operate and fully developing an incident response plan. In response to the report, the Bureau indicated it would work to identify the best way to address our recommendations. The Bureau reported that it has efforts under way to address our recommendations; however, more work remains. For example, according to Bureau officials they have been working to better track assessments of their common controls as part of a new risk management process. They expect to complete the transition to the new process by the end of this month. While the Bureau has recently taken key steps to address its IT security weaknesses, certain steps remain. Having adequate controls over access to its systems, as we recommended, would help the Bureau to better ensure that its information and systems are protected from intrusion.

CONCLUDING OBSERVATIONS

The Bureau is moving forward along a number of fronts to secure a more cost-effective 2020 enumeration. Significant research is already under way, and the Bureau is responding to our past recommendations. A little more than 6 years remains until Census Day 2020. While this might seem like an ample amount of time to finalize the Bureau's planning process and take steps to control costs, past experience has shown that the

---

[21] GAO, *Information Security: Actions Needed by Census Bureau to Address Weaknesses*, GAO-13-63 (Washington, D.C.: Jan. 22, 2013).

chain of interrelated preparations that need to occur at specific times and in the right sequence leave little room for delay or missteps.

Thus, as the Bureau's 2020 planning and reform efforts gather momentum, the effectiveness of those efforts will be determined in large measure by the extent to which they enhance the Bureau's ability to control costs, ensure quality, and adapt to future technological and societal changes. Likewise, Congress can hold the Bureau accountable for results, weigh in on key design decisions, provide the Bureau with resources the Congress believes are appropriate to support that design, and help ensure that the gains made to date stay on track. The Bureau's initial preparations for 2020 are making progress. Nonetheless, continuing congressional oversight remains vital.

Chairman Farenthold, Ranking Member Lynch, and Members of the Subcommittee, this concludes our statement today. We would be pleased to respond to any questions that you may have.

In: Actions Needed ...
Editor: Bryant Schneider
ISBN: 978-1-53616-716-0
© 2019 Nova Science Publishers, Inc.

Chapter 4

2020 CENSUS: ACTIONS NEEDED TO MITIGATE KEY RISKS JEOPARDIZING A COST-EFFECTIVE AND SECURE ENUMERATION[*]

United States Government Accountability Office

WHY GAO DID THIS STUDY

One of the Bureau's most important functions is to conduct a complete and accurate decennial census of the U.S. population. The decennial census is mandated by the Constitution and provides vital data for the nation. A complete count of the nation's population is an enormous undertaking as the Bureau seeks to control the cost of the census, implement operational innovations, and use new and modified IT systems. In recent years, GAO

---

[*] This is an edited, reformatted and augmented version of United States Government Accountability Office; Report to Congressional Requesters, Publication No. GAO-18-543T, dated May 2018.

has identified challenges that raise serious concerns about the Bureau's ability to conduct a cost-effective count. For these reasons, GAO added the 2020 Census to its high-risk list in February 2017.

GAO was asked to testify about the Bureau's progress in preparing for the 2020 Census. To do so, GAO summarized its prior work regarding the Bureau's planning efforts for the 2020 Census. GAO also included preliminary observations from its ongoing work examining the 2018 End-to-End Test. This information is related to, among other things, progress on key systems to be used for the 2018 End-to-End Test, including the status of IT security assessments, and efforts to update the life-cycle cost estimate.

WHAT GAO RECOMMENDS

Over the past decade, GAO has made 84 recommendations specific to the 2020 Census to address the issues raised in this and other products. The Bureau generally has agreed with the recommendations. As of May 2018, 30 recommendations had not been fully implemented.

WHAT GAO FOUND

The Census Bureau (Bureau) is planning several innovations for the 2020 Decennial Census, including re-engineering field operations by relying more on automation, using administrative records such as Medicare and Medicaid records, to supplement census data, verifying addresses in-office using on- screen imagery, and allowing the public to respond using the Internet. These innovations show promise for controlling costs, but they also introduce new risks, in part, because they have not been used extensively in earlier enumerations, if at all. As a result, robust testing is needed to ensure that key systems and operations will function as planned. However, citing budgetary uncertainties, the Bureau canceled its 2017 field test and then scaled back its 2018 End-to End Test by reducing the number of test sites from three to one. Without sufficient testing, operational problems can go undiscovered and the opportunity to improve operations will be lost, as key census-taking activities will not be tested across a range of geographic locations, housing types, and demographic groups.

The Bureau continues to face challenges in managing and overseeing the information technology (IT) programs, systems, and contracts supporting the 2020 Census. For example, GAO's ongoing work has determined that the schedule for developing IT systems to support the 2018 End-to-End Test has experienced several delays. Further, the Bureau has not addressed several security risks and challenges to its systems and data, including making certain that security assessments are completed in a timely manner, and that risks are at an acceptable level. Given that operations for the 2018 End-to-End Test began in August 2017, it is important that the Bureau quickly address these challenges.

In addition, the Bureau needs to control any further cost growth and develop cost estimates that reflect best practices. In October 2017, the Department of Commerce announced that it had updated its October 2015 life-cycle cost estimate and now projects the life-cycle cost of the 2020 Census will be $15.6 billion, a more than $3 billion (27 percent) increase over its earlier estimate (see figure). The higher estimated life-cycle cost is due, in part, to the Bureau's earlier failure to meet best practices for a quality cost estimate.

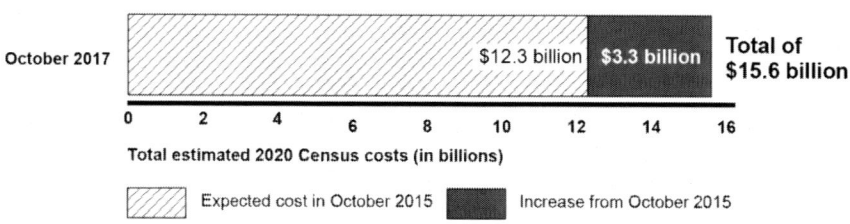

Source: GAO analysis of Census Bureau data. | Gao-18-543T

Increases to the 2020 Census Life-Cycle Costs Estimated by the Census Bureau.

The Bureau provided GAO with the documentation used to develop the $15.6 billion cost estimate. Based on its preliminary analysis, GAO has found that the Bureau has made improvements in its cost estimation process across the best practices.

Chairman Gowdy, Ranking Member Cummings, and Members of the Committee:

We are pleased to be here today to discuss the U.S. Census Bureau's (Bureau) progress in preparing for the 2020 Decennial Census. Conducting the decennial census of the U.S. population is mandated by the Constitution and provides vital data for the nation. The information that the census collects is used to apportion the seats of the House of Representatives; redraw congressional districts; allocate billions of dollars each year in federal financial assistance; and provide a social, demographic, and economic profile of the nation's people to guide policy decisions at each level of government. Further, businesses use census data to market new services and products and to tailor existing ones to demographic changes.

For 2020, a complete count of the nation's population is an enormous undertaking. The Bureau, a component of the Department of Commerce (Commerce), is seeking to control the cost of the census while it implements several innovations and manages the processes of acquiring and developing information technology (IT) systems. In recent years, we have identified challenges that raise serious concerns about the Bureau's ability to conduct a cost-effective count of the nation, including issues with the agency's research, testing, planning, scheduling, cost estimation, systems development, and IT security practices.

Over the past decade, we have made 84 recommendations specific to the 2020 Census to help address these and other issues. Commerce has generally agreed with those recommendations and has made progress in implementing them. However, 30 of our recommendations had not been fully implemented as of May 2018, although the Bureau had taken initial steps to implement many of them.

We also added the 2020 Decennial Census to GAO's high-risk list in February 2017.[1] As preparations for the next census ramp up, fully implementing our recommendations to address the risks jeopardizing the 2020 Census is more critical than ever.

Currently, the Bureau is conducting the 2018 End-to-End Test, which began in August 2017 and runs through April 2019. This effort is the Bureau's final opportunity to test all key systems and operations to ensure readiness for the 2020 Census.

Our testimony today will describe (1) why we added the decennial census to our high risk list, including challenges in implementing and securing IT systems; and (2) the steps that Commerce and the Bureau can take going forward to mitigate the risks jeopardizing a cost-effective census.

The information in this statement is based primarily on prior work regarding the Bureau's planning efforts for 2020.[2] For that body of work, we reviewed, among other things, relevant Bureau documentation, including the 2020 Census Operational Plan; recent decisions on preparations for the 2020 Census; and outcomes of key IT milestone reviews. We also discussed the status of our recommendations with Commerce and Bureau staff. Other details on the scope and methodology

---

[1] GAO, *High-Risk Series: Progress on Many High-Risk Areas, While Substantial Efforts Needed on Others*, GAO-17-317 (Washington, D.C.: Feb. 15, 2017). GAO maintains a high-risk program to focus attention on government operations that it identifies as high risk due to their greater vulnerabilities to fraud, waste, abuse, and mismanagement or the need for transformation to address economy, efficiency, or effectiveness challenges.

[2] For example, GAO, *2020 Census: Actions Needed to Mitigate Key Risks Jeopardizing a Cost-Effective Enumeration*, GAO-18-215T (Washington, D.C.: Oct. 31, 2017); *2020 Census: Continued Management Attention Needed to Oversee Innovations, Develop and Secure IT Systems, and Improve Cost Estimation*, GAO-18-141T (Washington, D.C.: Oct. 12, 2017); *2020 Census: Bureau Is Taking Steps to Address Limitations of Administrative Records*, GAO-17-664 (Washington, D.C.: July 26, 2017); *2020 Census: Bureau Needs to Better Leverage Information to Achieve Goals of Reengineered Address Canvassing*, GAO-17-622 (Washington, D.C.: July 20, 2017); *2020 Census: Sustained Attention to Innovations, IT Systems, and Cost Estimation Is Needed*, GAO-17-584T (Washington, D.C.: May 3, 2017); *2020 Census: Additional Actions Could Strengthen Field Data Collection Efforts*, GAO-17-191 (Washington, D.C.: Jan. 26, 2017); *Information Technology: Better Management of Interdependencies between Programs Supporting 2020 Census Is Needed*, GAO-16-623 (Washington, D.C.: Aug. 9, 2016); *2020 Census: Census Bureau Needs to Improve Its Life-Cycle Cost Estimating Process*, GAO-16-628 (Washington, D.C.: June 30, 2016); and *2020 Census: Additional Actions Would Help the Bureau Realize Potential Administrative Records Cost Savings*, GAO-16-48 (Washington, D.C.: Oct. 20, 2015).

for our prior work are provided in each published report on which this testimony is based.

In addition, we include information in this statement from our ongoing work on the readiness of the Bureau's IT systems for the 2018 End-to- End Test. Specifically, we collected and reviewed documentation on the status and plans for system development, testing, and security assessments for the 2018 End-to-End Test. This includes the Bureau's integration and implementation plan, solution architecture, and memorandums documenting outcomes of security assessments. We also interviewed relevant agency officials.

We provided the information in this statement to the Bureau for comment on April 30, 2018. The Bureau provided technical comments, which we addressed as appropriate.

We conducted the work on which this statement is based in accordance with generally accepted government auditing standards. Those standards require that we plan and perform the audit to obtain sufficient, appropriate evidence to provide a reasonable basis for our findings and conclusions based on our audit objectives. We believe that the evidence obtained provides a reasonable basis for our findings and conclusions based on our audit objectives.

BACKGROUND

The cost of counting the nation's population has been escalating with each decade. The 2010 Census was the most expensive in U.S. history at about $12.3 billion, and was about 31 percent more costly than the $9.4 billion 2000 Census (in 2020 constant dollars).[3] According to the Bureau, the total cost of the 2020 Census is now estimated to be approximately $15.6 billion dollars, more than $3 billion higher than previously estimated by the Bureau.

---

[3] According to the Bureau, these figures rely on fiscal year 2020 constant dollar factors derived from the Chained Price Index from "Gross Domestic Product and Deflators Used in the Historical Tables: 1940–2020" table from the Fiscal Year 2016 Budget of the United States Government.

Moreover, as shown in Figure 1, the average cost for counting a housing unit increased from about $16 in 1970 to around $92 in 2010 (in 2020 constant dollars). At the same time, the return of census questionnaires by mail (the primary mode of data collection) declined over this period from 78 percent in 1970 to 63 percent in 2010. Declining mail response rates have led to higher costs because the Bureau sends temporary workers to each non-responding household to obtain census data.

Achieving a complete and accurate census has become an increasingly daunting task, in part, because the population is growing larger, more diverse, and more reluctant to participate in the enumeration. In many ways, the Bureau has had to invest substantially more resources each decade to conduct the enumeration.

In addition to these external societal challenges that make achieving a complete count a daunting task, the Bureau also faces a number of internal management challenges that affect its capacity and readiness to conduct a cost-effective enumeration. Some of these issues—such as acquiring and developing IT systems and preparing reliable cost estimates—are long-standing in nature.

At the same time, as the Bureau looks toward 2020, it also faces newly emerging and evolving uncertainties. For example, on March 26, 2018, the Secretary of Commerce announced his decision to add a question to the decennial census on citizenship status. In our prior work we have noted the risks associated with late changes of any nature to the design of the census if the Bureau is unable to fully test those changes under operational conditions.[4]

The Bureau also faced budgetary uncertainties that, according to the Bureau, led to the curtailment of testing in 2017 and 2018. However, the Consolidated Appropriations Act, 2018 appropriated for the Periodic Censuses and Programs account $2.544 billion, which more than doubles the Bureau's request in the President's Fiscal Year 2018 Budget of $1.251

---

[4] GAO, *2010 Census: Little Time Remains to Address Operational Challenges*, GAO-09-408T (Washington, D.C.: Mar. 5, 2009).

billion.[5] According to the explanatory statement accompanying the act, the appropriation, which is available through fiscal year 2020, is provided to ensure the Bureau has the necessary resources to immediately address any issues discovered during the 2018 End-to-End Test, and to provide a smoother transition between fiscal year 2018 and fiscal year 2019.[6]

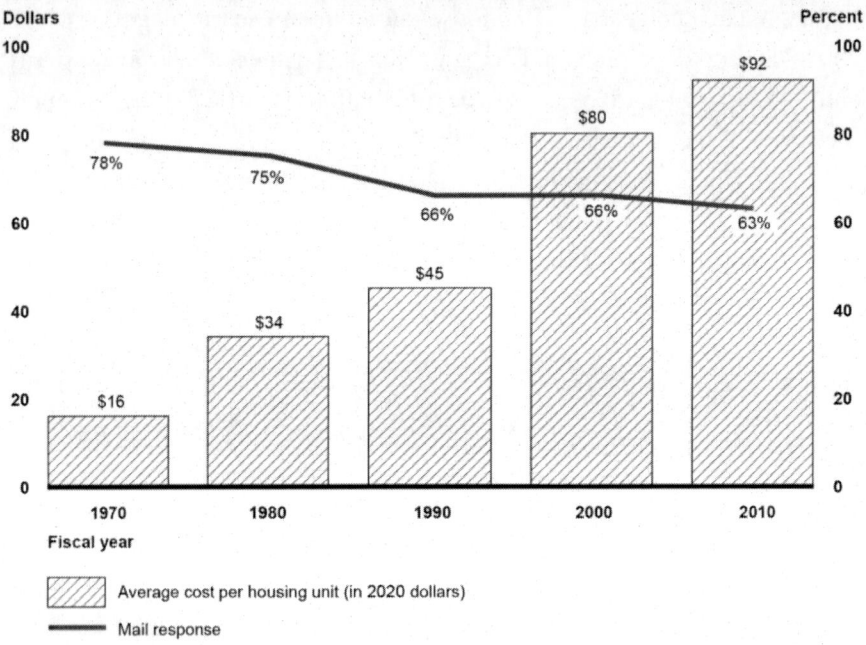

Source: GAO analysis of Census Bureau data. | Gao-18-543T

Figure 1. The Average Cost of Counting Each Housing Unit (in 2020 Constant Dollars) Has Escalated Each Decade, While the Percentage of Mail Response Rates Has Declined.

---

[5] Consolidated Appropriations Act, 2018, Pub. L. No. 115-141, Division B, Title I (Mar. 23, 2018).

[6] Joint explanatory statement of conference, 164 Cong. Rec. H2045, H2084 (daily ed. Mar. 22, 2018) (statement of Chairman Frelinghuysen), specifically referenced in section 4 of the Consolidated Appropriations Act, 2018, Pub. L. No. 115-141, § 4 (Mar. 23, 2018).

Moreover, according to Bureau officials, this level of funding helps the Bureau to complete testing and carry out other activities as planned.

The Bureau Plans to Rely Heavily on IT for the 2018 End-to-End Test and the 2020 Census

The Bureau plans to rely heavily on both new and legacy IT systems and infrastructure to support the 2018 End-to-End Test and the 2020 Census operations. For example, the Bureau plans to deploy and use 44 systems in the 2018 End-to-End Test.[7] Eleven of these systems are currently being developed or modified as part of an enterprise-wide initiative called Census Enterprise Data Collection and Processing (CEDCaP).[8] This initiative is a large and complex modernization program intended to deliver a system-of-systems to support all of the Bureau's survey data collection and processing functions, rather than continuing to rely on unique, survey-specific systems with redundant capabilities.[9]

To support the 2018 End-to-End Test, the Bureau plans to incrementally deploy and use the 44 systems from December 2016 through the end of the test in April 2019. These systems are to be used to support the operations involved in the test, including address canvassing,[10] self-response (i.e., Internet, phone, or paper), field enumeration, and tabulation and dissemination.

---

[7] In October 2017, we reported that the Bureau planned to use 43 systems in the 2018 End-to-End Test. Since that time, the Bureau has added 3 systems to the system list for the 2018 End-to-End Test and removed 2 systems. As of March 2018, the Bureau plans to use 52 systems during 2020 Census operations.

[8] The Bureau is pursuing enterprise-wide technology solutions intended to support other major surveys the Bureau conducts as well, such as the American Community Survey and the Economic Census.

[9] As a result of the Bureau's challenges in implementing key IT internal controls and its rapidly approaching deadline for conducting the decennial census, we identified CEDCaP as an IT investment in need of attention in both our February 2015 and February 2017 high-risk reports.

[10] The purpose of address canvassing is to deliver a complete and accurate address list and maps for enumeration purposes.

KEY RISKS ARE JEOPARDIZING A COST-EFFECTIVE ENUMERATION

We added the 2020 Census to our list of high-risk programs in February 2017 because (1) innovations never before used in prior enumerations are not expected to be fully tested, (2) the Bureau continues to face challenges in implementing and securing IT systems, and (3) the Bureau's October 2015 cost estimate was unreliable.[11] If not sufficiently addressed, these risks could adversely impact the cost and quality of the enumeration. Moreover, the risks are compounded by other factors that contribute to the challenge of conducting a successful census, such as the nation's increasingly diverse population and concerns over personal privacy.

Key Risk #1: The Bureau Has Redesigned the Census with the Intent to Control Costs, but has Scaled Back Critical Tests

The basic design of the enumeration—mail out and mail back of the census questionnaire with in-person follow-up for non-respondents—has been in use since 1970. However, a lesson learned from the 2010 Census and earlier enumerations is that this traditional design is no longer capable of cost-effectively counting the population.

In response to its own assessments, our recommendations, and studies by other organizations, the Bureau has fundamentally re-examined its approach for conducting the 2020 Census. Specifically, its plan for 2020 includes four broad innovation areas: re-engineering field operations, using administrative records, verifying addresses in-office, and developing an Internet self-response option (see Table 1).

If they function as planned, the Bureau initially estimated that these innovations could result in savings of over $5 billion (in 2020 constant dollars) when compared to its estimates of the cost for conducting the

---

[11] GAO-17-317.

census with traditional methods. However, in June 2016, we reported that the Bureau's initial life-cycle cost estimate developed in October 2015 was not reliable and did not adequately account for risk.[12] As discussed earlier in this statement, the Bureau has updated its estimate from $12.3 billion and now estimates a life-cycle cost of $15.6 billion, which would result in a smaller potential savings from the innovative design than the Bureau originally estimated.[13] According to the Bureau, the goal of the cost estimate increase was to ensure quality was fully addressed.

Table 1. The Census Bureau (Bureau) Is Introducing Four Innovation Areas for the 2020 Census

Innovation area	Description
Re-engineered field operations	The Bureau intends to automate data collection methods, including its case management system.
Administrative records	In certain instances, the Bureau plans to reduce enumerator collection of data by using administrative records (information already provided to federal and state governments as they administer other programs such as, Medicare and Medicaid records).
Verifying addresses in-office	To ensure the accuracy of its address list, the Bureau intends to use "in-office" procedures and on-screen imagery to verify addresses and reduce street-by-street field canvassing.
Internet self-response option	The Bureau plans to offer households the option of responding to the survey through the Internet. The Bureau has not previously offered such an option on a large scale.

Source: GAO analysis of Census Bureau data. | GAO-18-543T

While the planned innovations could help control costs, they also introduce new risks, in part, because they include new procedures and technology that have not been used extensively in earlier decennials, if at all. Our prior work has shown the importance of the Bureau conducting a robust testing program, including the 2018 End-to-End Test.[14] Rigorous

---

[12] GAO-16-628.

[13] The historical life-cycle cost figures for prior decennials as well as the initial estimate for 2020 provided by Commerce in October 2017 differ slightly from those reported by the Bureau previously. According to Commerce documents, the more recently reported figures are "inflated to the current 2020 Census time frame (fiscal years 2012 to 2023)," rather than to 2020 constant dollars as the earlier figures had been. Specifically, since October 2017, Commerce and the Bureau have reported the October 2015 estimate for the 2020 Census as $12.3 billion; this is slightly different than the $12.5 billion the Bureau had initially reported.

[14] GAO-17-622.

testing is a critical risk mitigation strategy because it provides information on the feasibility and performance of individual census-taking activities, their potential for achieving desired results, and the extent to which they are able to function together under full operational conditions. To address some of these challenges we have made numerous recommendations aimed at improving reengineered field operations, using administrative records, verifying the accuracy of the address list, and securing census responses via the Internet.

The Bureau has held a series of operational tests since 2012, but according to the Bureau, it has scaled back recent tests because of funding uncertainties. For example, the Bureau canceled the field components of the 2017 Census Test including non-response follow-up, a key census operation.[15] In November 2016, we reported that the cancelation of the 2017 Census Test was a lost opportunity to test, refine, and integrate operations and systems, and that it put more pressure on the 2018 End-to-End Test to demonstrate that enumeration activities will function under census-like conditions as needed for 2020.

However, in May 2017, the Bureau scaled back the operational scope of the 2018 End-to-End Test and, of the three planned test sites; only the Rhode Island site would fully implement the 2018 End-to-End Test. The Washington and West Virginia sites would test just one field operation. In addition, due to budgetary concerns, the Bureau decided to remove three coverage measurement operations (and the technology that supports them) from the scope of the test.[16] However, removal of the coverage measurement operations does not affect delivery of apportionment or redistricting data.

Without sufficient testing, operational problems can go undiscovered and the opportunity to improve operations will be lost, in part because the 2018 End-to-End Test is the last opportunity to demonstrate census technology and procedures across a range of geographic locations, housing

---

[15] In non-response follow-up, if a household does not respond to the census by a certain date, the Bureau will conduct an in-person visit by an enumerator to collect census data using a mobile device provided by the Bureau.

[16] Coverage measurement evaluates the quality of the census data by estimating the census coverage based on a post-enumeration survey.

types, and demographic groups. We plan to issue a report in June 2018 on address canvassing at the three test sites.

Key Risk #2: The Bureau Continues to Face Challenges in Implementing and Securing IT Systems

We have previously reported that the Bureau faces challenges in managing and overseeing IT programs, systems, and contractors supporting the 2020 Census. Specifically, we have noted challenges in its efforts to manage the schedules, contracts, costs, governance and internal coordination, and security for its systems. As a result of these challenges, the Bureau is at risk of being unable to fully implement the systems necessary to support the 2020 Census and conduct a cost- effective enumeration. We previously recommended that the Bureau take action to improve its implementation and management of IT in areas such as governance and internal coordination.[17]

Schedule Management

Our ongoing work has determined that the Bureau faces significant challenges in managing its schedule for developing and testing systems for the 2018 End-to-End Test that began in August 2017. As of April 2018, 30 of the 44 systems in the test had completed all development activities, while the remaining 14 were in the process of completing these activities. Figure 2 summarizes the development status for the 44 systems planned for the 2018 End-to-End Test, as of April 2018.

In addition, as of April 2018, 8 of the 44 systems had completed all testing activities (e.g., system and integration testing) for the 2018 End-to-End Test, while the remaining 36 were in the process of completing these activities. Figure 3 summarizes the status of testing for the 44 systems in the 2018 End-to-End Test. In addition, appendix I includes additional details about the status of development and testing for these systems.

---

[17] GAO-16-623.

Nevertheless, significant development and testing work remains to be completed. As stated previously, the 44 systems in the test are to be deployed multiple times in a series of operations (such as field enumeration). As of April 2018, 40 of the 44 systems had deployed at least a portion of functionality to support operations that have already occurred. The remaining system development and testing work is needed to support the 2018 End-to-End Test operations that are in process or planned.

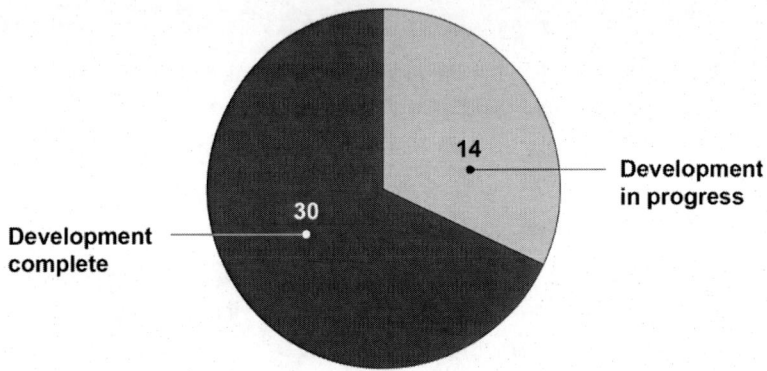

Source: GAO analysis of Census Bureau data. | GAO-18-543T

Figure 2. Development Status for the 44 Systems in the 2018 End-to-End Test, as of April 2018.

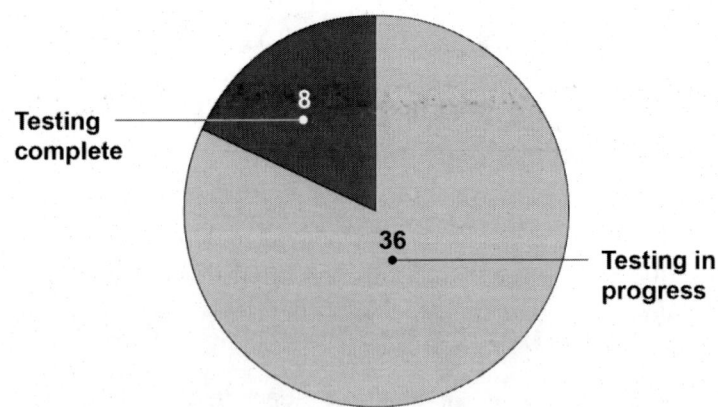

Source: GAO analysis of Census Bureau data. | GAO-18-543T

Figure 3. Testing Status for the 44 Systems in the 2018 End-to-End Test, as of April 2018.

2020 Census

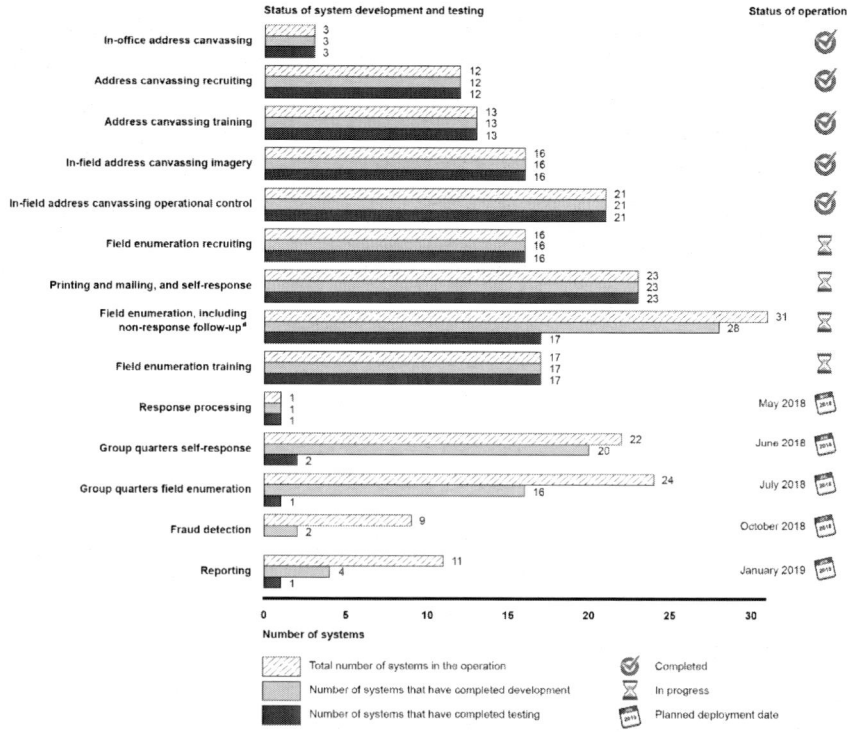

Source: GAO analysis of Census Bureau data. | GAO-18-543T

[a] In addition to non-response follow-up, field enumeration also includes operations such as update/leave and coverage improvement. In update/leave, listers update a housing unit's address and leave a questionnaire to allow the household to self-respond. The goal of coverage improvement is to resolve erroneous enumerations (such as people counted in the wrong place or more than once) and omissions.

Figure 4. Development and Testing Status for System Operations in the 2018 End-to-End Test, as of April 2018.

Specifically, as of April 2018, the Bureau had completed the development and testing for all of the systems supporting 9 of the 14 operations in the 2018 End-to-End Test, such as in-field address canvassing and response processing. However, the agency was in the process of completing system development and testing to support the remaining 5 operations. These 5 operations include 1 that is in progress—field enumeration—and 4 others that are planned for the future, including group quarters[18] enumeration and fraud detection. Figure 4 depicts the total number of systems supporting each operation, the number of systems that have completed development and testing, and the status of the operation.

However, due in part to challenges experienced during systems development, the Bureau has delayed by several months key IT milestone dates (e.g., dates to begin integration testing) for 9 of the 14 operations in the 2018 End-to-End Test. For example, the Bureau has moved the test readiness review date for the fraud detection operation from April 2018 to July 2018–a delay of 3 months—and the test readiness review of the group quarters field enumeration operation from October 2017 to April 2018—a delay of 6 months. Figure 5 depicts the delays in the key IT milestone dates for the operations in the 2018 End-to-End Test, as of April 2018. In total, since August 2017, the Bureau has delayed the final deployment date for 19 systems supporting the 14 operations.[19] Appendix I includes additional details about the delays in deployment dates since August 2017 for the 44 systems in the 2018 End- to-End Test.

We previously testified in May 2017 that the Bureau had faced similar challenges leading up to the 2017 Census Test, including experiencing delays in system development that led to compressed time frames for security reviews and approvals.[20] Specifically, we noted that the Bureau did not have time to thoroughly assess the low-impact components of one system and complete penetration testing[21] for another system prior to the test. Nonetheless, the Bureau's Chief Information Officer (CIO) accepted the security risks and uncertainty due to compressed time frames before the planned deployment. We highlighted that, for the 2018 End-to-End Test, it is important that these security assessments be completed in a timely

---

[18] Group quarters refer to college dormitories, nursing homes, and other facilities typically owned or managed by an entity providing housing, services, or both for the residents.

[19] According to officials within the Bureau's 2020 Census Systems Engineering and Integration office, the delays in the final deployment date for these systems are due to changes in the timing of the operations that the systems are supporting for the 2018 End- to-End Test.

[20] GAO-17-584T.

[21] The National Institute of Standards and Technology defines penetration testing as security testing in which evaluators mimic real-world attacks in an attempt to identify ways to circumvent the security features of an application, system, or network. Penetration testing often involves issuing real attacks on real systems and data, using the same tools and techniques used by actual attackers.

manner and that risks be at an acceptable level before the systems are deployed.

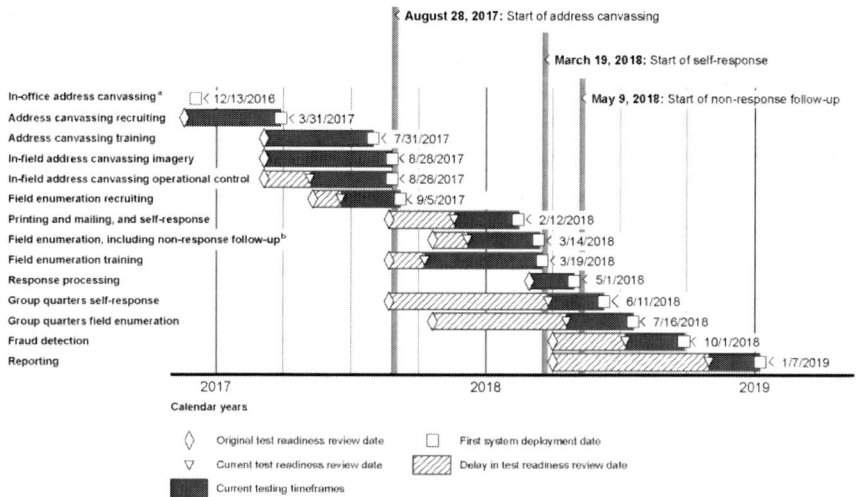

Source: GAO analysis of Census Bureau data. | GAO-18-543T

[a] The systems that supported the in-office address canvassing operation were in operations and maintenance and did not require additional testing before beginning in-office address canvassing.

[b] In addition to non-response follow-up, field enumeration also includes operations such as update/leave and coverage improvement. In update/leave, listers update a housing unit's address and leave a questionnaire to allow the household to self-respond. The goal of coverage improvement is to resolve erroneous enumerations (such as people counted in the wrong place or more than once) and omissions.

Figure 5. Delays in Key Information Technology Milestone Dates for System Operations in the 2018 End-to-End Test, as of April 2018.

As a result of the delays in system development and testing, the Bureau has had (and likely will continue to have) reduced time available to conduct the security reviews and approvals for the systems being used in the 2018 End-to-End Test. Officials in the Bureau's Office of Information Security stated that the original plan was to have at least 6 to 8 weeks to perform security assessments for each system.

However, given the compressed time frames, Bureau officials informed us that, in some instances, they have had 5 to 8 days to complete certain assessments. As a result, the security of all system components may not be assessed before deployment. According to the Bureau's Chief Information Security Officer, components that do not have all controls assessed are to be tracked until the assessments are completed, even if it is after the system deploys.

If the Bureau continues to experience delays in meeting development and testing milestones for the 2018 End-to-End Test, it may not be able to fully test production-level systems and operations in a census-like environment prior to the 2020 Census. As stated earlier, without sufficient testing, operational problems can go undiscovered and the opportunity to improve operations will be lost.

Additional System Development and Testing Planned After the 2018 End-to-End Test

After the 2018 End-to-End Test, the Bureau still has additional system development and testing activities planned leading up to the 2020 Census. The Bureau plans to use a total of 52 systems in 2020 Census operations, including the 44 systems that are planned to be used in the 2018 End-to-End Test, and 8 additional systems that were not included in the test.[22] The Bureau expects that the systems used in the 2018 End-to- End Test will need additional development and testing due to, among other things, new functionality to be added, the need to scale system performance for the number of respondents expected during the 2020 Census, or to address system defects identified during the 2018 End-to- End Test.

Following the 2018 End-to-End Test, the Bureau plans to develop, test, and deploy the 52 systems for the 2020 Census in four groups, or operational releases: (1) recruiting and hiring; (2) address canvassing; (3) self-response, non-response follow-up, and fraud detection; and (4)

---

[22] Several of these systems are for the coverage measurement operation, which was cut from the scope of the 2018 End-to-End Test. As stated previously, coverage measurement evaluates the quality of the census data by estimating the census coverage based on a post-enumeration survey.

reporting and coverage measurement.[23] The systems are generally grouped by the operation(s) they support in the 2020 Census. For example, the third operational release—which includes the most systems—has 47 systems to be used for self-response (including via the Internet), non-response follow-up, and fraud detection. These systems are expected to be deployed in November 2019 so that they will be ready for the Internet self-response operation, which begins in March 2020. Table 2 identifies the key dates for the operational releases for systems in the 2020 Census.

As planning for the 2020 Census continues, it will be important for the Bureau to provide adequate time for system development and testing activities. This will help ensure that the time available for security assessments is not reduced as it was in the 2017 Test, and as it has been, thus far, during the 2018 End-to-End Test. Without adequate time for completing these security assessments, the Bureau will be challenged in ensuring that risks are at an acceptable level before the systems are deployed for the 2020 Census.

Contract Management

Our ongoing work has also determined that the Bureau faces challenges in managing its significant contractor support. The Bureau is relying on contractor support in many areas to prepare for the 2020 Census. For example, it is relying on contractors to develop a number of systems and components of the IT infrastructure. These activities include (1) developing the IT platform (as part of the CEDCaP program) that is intended to be used to collect data from households responding via the Internet and telephone, and for non-response follow-up activities; (2) procuring the mobile devices and cellular service to be used for non-response follow-up;[24] and (3) deploying the IT and telecommunications hardware in the field offices. According to Bureau officials, contractors are

---

[23] Similar to the 2018 End-to-End Test, a system being used in the 2020 Census may be deployed multiple times (with additional or new functionality) if that system is needed for more than one of these operations.

[24] In non-response follow-up, if a household does not respond to the census by a certain date, the Bureau will send out employees to visit the home. The Bureau's plan is for these enumerators to use a census application, on a mobile device provided by the Bureau, to capture the information given to them by the in-person interviews.

also providing support in areas such as fraud detection, cloud computing services, and disaster recovery.

Table 2. The Census Bureau's Planned Operational Releases for the 2020 Census, as of April 2018

Operational release name	Number of systems in the operational release	Expected completion date for system development	Expected completion date for integration and test	Expected deployment date
1. Recruiting and hiring	19	May 2018	July 2018	September 2018
2. Address canvassing	26	November 2018	March 2019	May 2019
3. Self-response, non-response follow-up, and fraud detection	47	February 2019	June 2019	November 2019
4. Reporting and coverage measurement	22	October 2019	February 2020	July 2020

Source: GAO analysis of Census Bureau data. | GAO-18-543T

In addition to the development of technology, the Bureau is relying on a technical integration contractor to integrate all of the key systems and infrastructure. The Bureau awarded a contract to integrate the 2020 Census systems and infrastructure in August 2016. The contractor's work was to include evaluating the systems and infrastructure and acquiring the infrastructure (e.g., cloud or data center) to meet the Bureau's scalability and performance needs. It was also to include integrating all of the systems, supporting technical testing activities, and developing plans for ensuring the continuity of operations. Since the contract was awarded, the Bureau has modified the scope to also include assisting with operational testing activities, conducting performance testing for two Internet self-response systems, and providing technical support for the implementation of the paper data capture system.

However, our ongoing work has determined that the Bureau is facing staffing challenges that could impact its ability to manage and oversee the technical integration contractor. Specifically, the Bureau is managing the integration contractor through a government program management office, but this office is still filling vacancies. As of February 2018, the Bureau reported that 34 of the office's 58 federal employee positions were vacant.

As a result, this program management office may not be sufficiently staffed to provide adequate oversight of contractor cost, schedule, and performance.

The development and testing schedule delays during the 2017 Test and preparations for the 2018 End-to-End Test raise concerns about the Bureau's ability to effectively perform contractor management. As we reported in November 2016, a greater reliance on contractors for these components of the 2020 Census requires the Bureau to focus on sound management and oversight of the key contracts, projects, and systems.[25] We will continue to monitor the Bureau's contract management as part of our ongoing work.

IT Cost Growth

The Bureau faces challenges in controlling IT cost growth. Specifically, the Bureau's October 2015 cost estimate included about $3.41 billion in total IT costs for fiscal years 2012 through 2023. These included costs for, among other things, system engineering, test and evaluation, and infrastructure, as well as for a portion of the CEDCaP program.[26] However, in October 2017, we reported[27] that IT costs would likely be at least $4.8 billion due to increases in costs associated with the CEDCaP program[28] and certain IT contracts (including those associated with technical integration and mobile devices).

In December 2017, the Bureau reported that its estimated IT costs had grown from $3.41 billion to $4.97 billion—an increase of $1.56 billion. Figure 6 identifies the Bureau estimate of total IT costs associated with the 2020 program as of December 2017.

---

[25] GAO, *Information Technology: Uncertainty Remains about the Bureau's Readiness for a Key Decennial Census Test*, GAO-17-221T (Washington, D.C.: Nov. 16, 2016).

[26] The 2020 program pays for a portion of the costs for the CEDCaP program. According to the October 2015 estimate, the portion of CEDCaP costs associated with the 2020 Census was estimated at $328 million of the $548 million total program estimate.

[27] GAO-18-215T.

[28] In May 2017, the Bureau reported that the CEDCaP program's cost estimate was increasing by about $400 million—from its original estimate of $548 million in 2013 to a revised estimate of $965 million in May 2017.

The cost increases were due, in large part, to the Bureau (1) updating the cost estimate for the CEDCaP program, (2) including an estimate for technical integration services, and (3) updating costs related to other major contracts (such as mobile device as a service).[29] Table 3 describes the IT costs that comprised the Bureau's cost estimate as of December 2017.

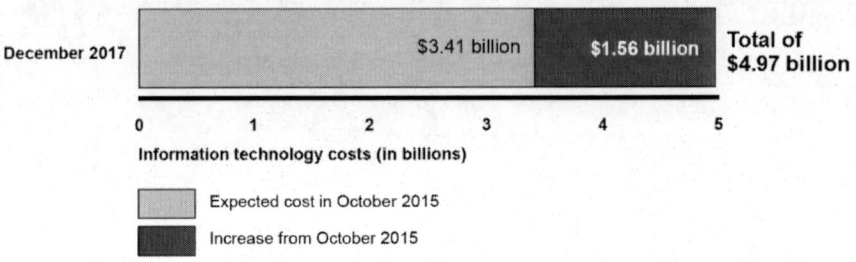

Source: GAO analysis of Census Bureau data. | GAO-18-543T

Figure 6. Total Information Technology Costs Estimated by the Census Bureau, as of December 2017.

Table 3. Total 2020 Census Information Technology (IT) Costs Estimated by the Census Bureau, by Cost Category, as of December 2017

IT cost category	Expected cost (in millions)
Technical integration services	$1,492
Census questionnaire assistance	$817
Other IT services, such as day-to-day IT support	$808
Census Enterprise Data Collection and Processing (CEDCaP) costs related to the 2020 Census	$509[a]
Decennial device as a service	$489
Field IT deployment	$450
Other non-CEDCaP systems, such as recruitment and personnel systems	$401
Total	$4,966

Source: GAO analysis of Census Bureau data. | GAO-18-543T

[a] The 2020 program pays for a portion of the costs for the CEDCaP program. As of May 2017, the Census Bureau estimated that the entire cost of the CEDCaP program would be about $965 million.

---

[29] As part of mobile device as a service, the Bureau plans to provide mobile devices (including mobile phones, tablets, and laptops) and cellular service to field staff to support operations such as address canvassing and non-response follow-up.

IT cost information that is accurately reported and clearly communicated is necessary to help ensure that Congress and the public have confidence that taxpayer funds are being spent in an appropriate manner. However, the amount of cost growth since the October 2015 estimate raises questions as to whether the Bureau has a complete understanding of the IT costs associated with the 2020 program. In December 2017, the Bureau provided us with the documentation used to develop the updated cost-estimate, and we are reviewing these materials as part of our continuing body of work on the 2020 Census.

Governance and Internal Coordination

Effective governance can drive change, provide oversight, and ensure accountability for results. Further, effective IT governance was envisioned in the statutory provisions enacted in 2014 and referred to as the Federal Information Technology Acquisition Reform Act (FITARA),[30] which strengthened and reinforced the role of the departmental CIO. The component CIO (such as the Bureau's CIO) also plays a role in effective IT governance, as the component is subject to the oversight and policies of the parent department implementing FITARA.

Our ongoing work has determined that officials in Commerce's Office of the Secretary have increased their oversight of the Bureau's preparations for the 2020 Census by holding regular meetings to discuss contracts, expected costs, and risks, among other topics. For example, Bureau officials reported that they have recently begun meeting with the Secretary of Commerce on a monthly basis and with the Under Secretary of Commerce for Economic Affairs on a weekly basis to discuss 2020 Census issues. Moreover, the department's Acting CIO has also been involved in overseeing the Bureau's IT system readiness. The Bureau has also appointed two new assistant directors within the Decennial Directorate. Each of these individuals is responsible for overseeing aspects of the 2020 Census program, to include schedules, contracts, and system development.

---

[30] Carl Levin and Howard P. 'Buck' McKeon National Defense Authorization Act for Fiscal Year 2015, Pub. L. No. 113-291, div. A, title VIII, subtitle D, 128 Stat. 3292, 3438-50 (Dec. 19, 2014).

In addition, to ensure executive-level oversight of the key systems and technology, the Bureau's CIO (or a designated representative) is to be a member of the governance boards that oversee all of the operations and technology for the 2020 Census. However, in August 2016 we reported on challenges that the Bureau has had with IT governance and internal coordination, including weaknesses in its ability to monitor and control IT project costs, schedules, and performance.[31] We made eight recommendations to the Secretary of Commerce to direct the Bureau to, among other things, better ensure that risks are adequately identified and schedules are aligned. Commerce agreed with our recommendations. As of May 2018, the Bureau had fully implemented five of the recommendations and had taken initial steps toward implementing the other three recommendations.

Further, given the schedule delays and cost increases previously mentioned, and the vast amount of development, testing, and security assessments left to be completed, we remain concerned about executive-level oversight of systems and security. Moving forward, it will be important that the CIO and other Bureau executives continue to use a collaborative governance approach to effectively manage risks and ensure that the IT solutions meet the needs of the agency within cost and schedule.

Information Security

In November 2016, we described the significant challenges that the Bureau faced in securing systems and data for the 2020 Census, and we noted that tight time frames could exacerbate these challenges.[32] Two such challenges were (1) ensuring that individuals gain only limited and appropriate access to the 2020 Census data, including personally identifiable information (PII), such as name, personal address, and date of birth; and (2) making certain that security assessments were completed in a timely manner and that risks were at an acceptable level.[33] Protecting PII, for example, is especially important because a majority of the 44 systems

---

[31] GAO-16-623.
[32] GAO-17-584T.
[33] GAO-17-221T.

to be used in the 2018 End-to-End Test contain such information, as reflected in Figure 7.[34]

To address these and other challenges, federal law specifies requirements for protecting federal information and information systems, such as those systems to be used in the 2020 Census. Specifically, the Federal Information Security Management Act of 2002 and the Federal Information Security Modernization Act of 2014 (FISMA) require executive branch agencies to develop, document, and implement an agency-wide program to provide security for the information and information systems that support operations and assets of the agency.[35]

Accordingly, the National Institute of Standards and Technology (NIST) developed risk management framework guidance for agencies to follow in developing information security programs.[36] In addition, the Office of Management and Budget's (OMB) revised Circular A-130 on managing federal information resources required agencies to implement the NIST risk management framework to integrate information security and risk management activities into the system development life cycle.[37]

In accordance with FISMA, NIST guidance, and OMB guidance, the Bureau's Office of the CIO established a risk management framework. This framework requires system developers to ensure that each of the Bureau's systems undergoes a full security assessment, and that system developers remediate critical deficiencies. In addition, according to the framework, system developers are to ensure that each component of a

---

[34] According to officials in the Bureau's Office of Information Security, 26 systems contain data that is protected from disclosure under Title 13 of the U.S. Code. This law protects information provided by the public for the Bureau's censuses and surveys and requires that the Bureau keep it confidential. 13 U.S.C. § 9. For example, the Bureau may not disclose or publish any private information that identifies an individual or business, such as names, addresses, Social Security numbers, and telephone numbers.

[35] The Federal Information Security Modernization Act of 2014, Pub. L. No. 113-283, 128 Stat. 3073 (Dec. 18, 2014) largely superseded the Federal Information Security Management Act of 2002, enacted as Title III, E-Government Act of 2002, Pub. L. No. 107-347, 116 Stat. 2899, 2946 (Dec. 17, 2002).

[36] NIST, *Guide for Applying the Risk Management Framework to Federal Information Systems: A Security Life Cycle Approach*, SP 800-37, Revision 1 (Gaithersburg, Md.: February 2010).

[37] OMB, *Revision of OMB Circular A-130, Managing Federal Information as a Strategic Resource* (Washington, D.C.: July 28, 2016).

system has its own system security plan that documents how the Bureau intends to implement security controls. As a result of this requirement, system developers for a single system might develop multiple system security plans which all have to be approved as part of the system's complete security documentation.

According to the Bureau's framework, each of the 44 systems in the 2018 End-to-End Test will need to have complete security documentation (such as system security plans) and an approved authorization to operate[38] prior to its use in the 2018 End-to-End Test. However, our ongoing work indicates that, while the Bureau is completing these steps for the 44 systems to be used in the 2018 End-to-End Test, significant work remains. Specifically:

- Six of the 44 systems are fully authorized to operate through the completion of the 2018 End-to-End Test.
- Thirty-two systems have a current authorization to operate, but the Bureau will need to reauthorize these systems before the completion of the 2018 End-to-End Test. Bureau officials in the CIO's Office of Information Security stated that these systems will need to be reauthorized because, among other things, they have additional development work planned that may require the systems to be reauthorized; are being moved to a different infrastructure environment (e.g., from a data center to a cloud-based environment); or have a current authorization that expires before the completion of the 2018 End-to-End Test.
- Six systems have not yet obtained an authorization to operate.

---

[38] According to the Bureau's framework, systems are to obtain security authorization approval from the authorizing official in order to operate. Specifically, the authorizing official evaluates the security authorization package and provides system authorization if the overall risk level is acceptable. In addition, according to the Bureau's IT security program policy, the issuance of an authorization to operate for a system requires support of both the technical authorizing official (i.e., the CIO) and the business authorizing official responsible for funding and managing the system (i.e., the Associate Director for Decennial Census Programs). Further, according to the Bureau's framework, once a system obtains an authorization, it is transitioned to the continuous monitoring process where the authorizing official can provide implicit, continued authorization for system operation as long as the risk level remains acceptable.

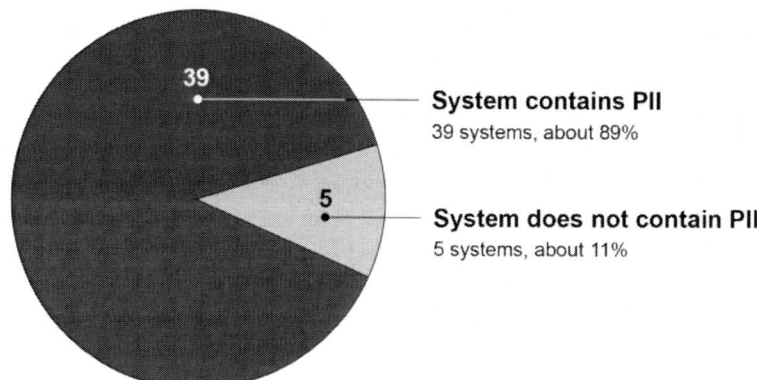

Source: GAO analysis of Census Bureau data. | GAO-18-543T

Figure 7. Personally Identifiable Information (PII) in Census Bureau Systems Included in the 2018 End-to-End Test, as of February 2018.

Figure 8 summarizes the authorization to operate status for the systems being used in the 2018 End-to-End Test, as reported by the Bureau.

Because many of the systems that will be a part of the 2018 End-to-End Test are not yet fully developed, the Bureau has not finalized all of the security controls to be implemented; assessed those controls; developed plans to remediate control weaknesses; and determined whether there is time to fully remediate any deficiencies before the systems are needed for the test. In addition, as discussed earlier, the Bureau is facing system development and testing challenges that are delaying the completion of milestones and compressing the time available for security testing activities.

Further, while the large-scale technological changes (such as Internet self-response) increase the likelihood of efficiency and effectiveness gains, they also introduce many information security challenges. The 2018 End-to-End Test also involves collecting PII on hundreds of thousands of households across the country, which further increases the need to properly secure these systems. Thus, it will be important that the Bureau provides adequate time to perform these security assessments, completes them in a timely manner, and ensures that risks are at an acceptable level before the systems are deployed.

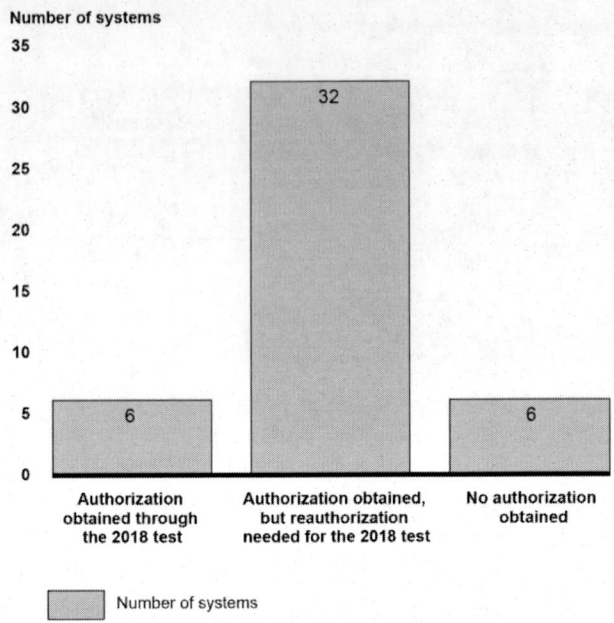

Source: GAO analysis of Census Bureau data. | GAO-18-543T

Figure 8. Authorization to Operate Status for 44 Systems Being Used in the Census Bureau's 2018 End-to-End Test, as of April 2018.

Key Risk #3: The Bureau will Need to Control Any Further Cost Growth and Develop Cost Estimates That Reflect Best Practices

In October 2017, Commerce announced that it had updated its October 2015 life-cycle cost estimate and now projects the life-cycle cost of the 2020 Census will be $15.6 billion, more than a $3 billion (27 percent) increase over the Bureau's earlier estimate. The higher estimated life-cycle cost is due, in part, as we reported in June 2016, to the Bureau's failure to meet best practices for a quality cost-estimate. Specifically, we reported that, although the Bureau had taken steps to improve its capacity to carry out an effective cost estimate, such as establishing an independent cost estimation office, its October 2015 version of the estimate for the 2020 Census only partially met two characteristics of a reliable cost estimate (comprehensive and accurate) and minimally met the other two (well-documented and credible). We also reported that risks were not properly accounted for in the cost estimate.[39]

We recommended that the Bureau take action to ensure its 2020 Census cost estimate meets all four characteristics of a reliable cost estimate, as well as properly account for risk to ensure there are appropriate levels for budgeted contingencies. The Bureau agreed with our recommendations. In response, Commerce reported that in May 2017, a multidisciplinary team was created to evaluate the 2020 Census program and to produce an independent cost estimate.

Factors driving the increased cost estimate include changes to assumptions relating to self-response rates, wage levels for temporary census workers, as well as the fact that major contracts and IT scale-up plans and procedures were not effectively planned, managed, and executed. The new estimate also includes a contingency of 10 percent of estimated costs per year as insurance against "unknown-unknowns," such as a major cybersecurity event.

The Bureau has provided us with the documentation used to develop the $15.6 billion cost estimate. We are reviewing these documents to determine the reliability of the estimate using our cost guide.[40] We plan to issue a report in the summer of 2018 on the revised estimate's reliability. In order for the new estimate to be deemed high quality, and, thus, the basis for any 2020 Census annual budgetary figures, the new cost-estimate will need to address the following four characteristics:

- Comprehensive. To be comprehensive an estimate should have enough detail to ensure that cost elements are neither omitted nor double-counted, and all cost-influencing assumptions are detailed in the estimate's documentation, among other things, according to best practices.

---

[39] GAO-16-628.
[40] GAO, *GAO Cost Estimating and Assessment Guide: Best Practices for Developing and Managing Capital Program Costs (Supersedes GAO-07-1134SP)*, GAO-09-3SP (Washington, D.C.: Mar. 2, 2009).

- Accurate. Accurate estimates are unbiased and contain few mathematical mistakes.
- Well-documented. Cost estimates are considered valid if they are well-documented to the point they can be easily repeated or updated and can be traced to original sources through auditing, according to best practices.
- Credible. Credible cost estimates must clearly identify limitations due to uncertainty or bias surrounding the data or assumptions, according to best practices.

Based on our preliminary analysis, we have found that the Bureau has made improvements in its cost estimation process across the four characteristics.

CONTINUED MANAGEMENT ATTENTION NEEDED TO KEEP PREPARATIONS ON TRACK AND HELP ENSURE A COST-EFFECTIVE ENUMERATION

2020 Challenges are Symptomatic of Deeper Long-Term Organizational Issues

The difficulties facing the Bureau's preparation for the decennial census in such areas as planning and testing; managing and overseeing IT programs, systems, and contractors supporting the enumeration; developing reliable cost estimates; prioritizing decisions; managing schedules; and other challenges, are symptomatic of deeper organizational issues.

Following the 2010 Census, a key lesson learned for 2020 that we identified was ensuring that the Bureau's organizational culture and structure, as well as its approach to strategic planning, human capital management, internal collaboration, knowledge sharing, capital decision-making, risk and change management, and other internal functions are aligned toward delivering more cost-effective outcomes.[41]

The Bureau has made improvements over the last decade, and continued progress will depend in part on sustaining efforts to strengthen risk management activities, enhancing systems testing, bringing in experienced personnel to key positions, implementing our recommendations, and meeting regularly with officials from its parent agency, Commerce. Going forward, we have reported that the key elements needed to make progress in high-risk areas are top-level attention by the administration and agency officials to (1) leadership commitment, (2) ensuring capacity, (3) developing a corrective action plan, (4) regular monitoring, and (5) demonstrated progress. Although important steps have been taken in at least some of these areas, overall, far more work is needed.[42] We discuss three of five areas below.

The Secretary of Commerce has taken several actions towards demonstrating leadership commitment. For example, the previously noted multidisciplinary review team included members with Bureau leadership experience, as well as members with private sector technology management experience. Additional program evaluation and the independent cost estimate was produced by a team from the Commerce Secretary's Office of Acquisition Management that included a member detailed from OMB. We have met with the Under Secretary of Commerce for Economic Affairs at her invitation to discuss oversight priorities, Commerce's commitment to them, as well as the progress on high risk areas. Commerce reports that other senior officials are now also actively involved in the management and oversight of the decennial. Likewise, with respect to monitoring progress, the Commerce Secretary reports having weekly 2020 Census oversight reviews with senior Bureau staff and will require metric tracking and program execution status on a real- time basis.

On the other hand, demonstrating the capacity to address high risk concerns remains a challenge. For example, as stated earlier, the Bureau is

---

[41] GAO, *2010 Census: Preliminary Lesson Learned Highlight the Need for Fundamental Reforms,* GAO-11-496T (Washington, D.C.: Apr. 6, 2011).

[42] GAO-17-317.

facing staffing challenges that could impact its ability to manage and oversee the technical integration contractor. Specifically, the Bureau is managing the integration contractor through a government program management office, but this office is still filling vacancies. As previously discussed, the Bureau reported that 34 of 58, or almost 60 percent, of the office's federal employee positions were vacant as of February 2018. As a result, this program management office may not be able to provide adequate oversight of contractor cost, schedule, and performance.

In the months ahead, we will continue to monitor the Bureau's progress in addressing in each of the five elements essential for reducing the risk to a cost-effective enumeration.

Strong Bureau Leadership Will Be Critical for Keeping Efforts On-Track

At a time when strong Bureau management is needed, vacancies in the agency's two top positions—Director and Deputy Director—are not helpful for keeping 2020 preparations on-track. These vacancies are due to the previous director's retirement on June 30, 2017, and the previous deputy director's appointment to be the Chief Statistician of the United States within OMB in January 2017. Although interim leadership has since been named, there are upcoming key decisions that would more effectively be made by permanent leadership. In our prior work we have noted how openings in the Bureau's top position makes it difficult to ensure accountability and continuity, as well as to develop and sustain efforts that foster change, produce results, mitigate risks, and control costs over the long term. For example, in September 2018 the Bureau is scheduled to decide what types of challenges it will allow state, tribal, or local governments to make to their official 2020 Census results.

The director of the Bureau is appointed by the President, by and with the advice and consent of the Senate, without regard to political affiliation.

The director's term is a fixed 5-year term of office.[43] An individual may be reappointed but may not serve more than two full terms as director. The director's position was first filled this way beginning on January 1, 2012, and cycles every fifth year thereafter. Because the new term began on January 1, 2017, the time that elapses until a new director is confirmed counts against the 5-year term of office. As a result, the next director's first term will be less than 5 years.

Going forward, filling these top two positions should be an important priority. On the basis of our prior work, key attributes of a census director, in addition to technical expertise and the ability to lead large, long-term and high risk programs, could include abilities in the following areas:

- Strategic Vision. The director needs to build a long-term vision for the Bureau that extends beyond the current decennial census. Strategic planning, human-capital succession planning, and life-cycle cost estimates for the Bureau all span the decade.
- Sustaining Stakeholder Relationships. The director needs to continually expand and develop working relationships and partnerships with governmental, political, and other professional officials in both the public and private sectors to obtain their input, support, and participation in the Bureau's activities.
- Accountability. The life-cycle cost for a decennial census spans a decade, and decisions made early in the decade about the next decennial census guide the research, investments, and tests carried out throughout the decennial census. Institutionalizing accountability over an extended period may help long-term decennial initiatives provide meaningful and sustainable results.[44]

---

[43] 13 U.S.C. § 21(a)-(b)(1).
[44] GAO, *2010 Census: Data Collection Operations Were Generally Completed as Planned, but Long-standing Challenges Suggest Need for Fundamental Reforms,* GAO-11-193 (Washington, D.C.: Dec. 14, 2010).

Further Actions Needed on Our Recommendations

Over the past several years we have issued numerous reports that underscored the fact that, if the Bureau was to successfully meet its cost savings goal for the 2020 Census, the Bureau needed to take significant actions to improve its research, testing, planning, scheduling, cost estimation, system development, and IT security practices. As previously stated, over the past decade, we have made 84 recommendations specific to the 2020 Census to help address these and other issues. The Bureau has generally agreed with those recommendations and has taken action to address them. However, 30 of the recommendations had not been fully implemented as of May 2018, although the Bureau had taken initial steps to implement many of them. We have designated 21 of these 84 recommendations as a priority for Commerce and 6 have been implemented. In April 2018, we sent the Secretary of Commerce a letter that identified our open priority recommendations at the department, 15 of which concern the 2020 Census.[45] We believe that attention to these recommendations is essential for a cost-effective enumeration. The recommendations included implementing reliable cost estimation and scheduling practices in order to establish better control over program costs, as well as taking steps to better position the Bureau to develop an Internet response option for the 2020 Census.

On October 3, 2017, in response to our August 2017 priority recommendation letter, the Commerce Secretary noted that he shared our concerns about the 2020 Census and acknowledged that some of the programs had not worked as planned, and are not delivering the savings that were promised. The Commerce Secretary also stated that he intends to improve the timeliness for implementing our recommendations.

We meet quarterly with Bureau officials to discuss the progress and status of open recommendations related to the 2020 Census, which has resulted in Bureau actions in recent months leading to closure of some recommendations. We are encouraged by this commitment by Commerce

---

[45] The 15 priority recommendations for the 2020 Census cover the period from November 2009 to July 2017.

and the Bureau in addressing our recommendations. Implementing our recommendations in a complete and timely manner is important because it could improve the management of the 2020 Census and help to mitigate continued risks.

In conclusion, while the Bureau has made progress in revamping its approach to the census, it faces considerable challenges and uncertainties in (1) implementing key cost-saving innovations and ensuring they function under operational conditions; (2) managing the development and security of IT systems; and (3) developing a quality cost estimate for the 2020 Census and preventing further cost increases. Without timely and appropriate actions, these challenges could adversely affect the cost, accuracy, schedule, and security of the enumeration. For these reasons, the 2020 Census is a GAO high risk area.

Going forward, continued management attention and oversight will be vital for ensuring that risks are managed, preparations stay on-track, and the Bureau is held accountable for implementing the enumeration, as planned. We will continue to assess the Bureau's efforts to conduct a cost-effective and secure enumeration and look forward to keeping Congress informed of the Bureau's progress.

Chairman Gowdy, Ranking Member Cummings, and Members of the Committee, this completes our prepared statement. We would be pleased to respond to any questions that you may have.

APPENDIX I: STATUS OF DEVELOPMENT AND TESTING FOR SYSTEMS IN THE 2018 END-TO-END TEST, AS OF APRIL 2018

As part of its 2018 End-to-End Test, the Census Bureau (Bureau) plans to deploy 44 systems incrementally to support key operations from December 2016 through the end of the test in April 2019. These operations include address canvassing, self-response (i.e., Internet, phone, or paper), field enumeration, and tabulation and dissemination.

Table 1. Development, Testing, and Deployment Status for the 44 Systems in the Census Bureau's 2018 End-to-End Test, as of April 2018

System name and description	Status of development	Status of testing	Expected/ actual first deployment date[a] (delay since August 2017)	Expected/ actual final deployment date[a] (delay since August 2017)	Has at least a portion of the system's functionality been deployed?
1. Block Assessment, Review and Classification Application Interactive review tool that is designed to assist an analyst in assessing a set of geographic work units.	Complete	Complete	December 2016	n/a	Yes
2. OneForm Designer Plus Tool that creates paper forms including decennial questionnaires, letters, envelopes, notices of visit, language guides, and other Decennial field and public materials.	Complete	Complete	March 2017	n/a	Yes
3. Census Document System Web-based system for requesting forms design services, publications and graphics services, and printing services.	Complete	Complete	March 2017	September 2017	Yes
4. MOJO Recruiting Dashboard System that provides a dashboard to show recruiting metrics.	Complete	Complete	March 2017	June 2018	Yes
5. Matching and Coding Software System that allows for clerical matching and geocoding during Non-ID Processing.	Complete	Complete	February 2018	n/a	Yes
6. Real Time Non-ID Processing System that matches addresses in real-time, geocodes addresses in real-time, and geo-locates housing units using web map services.	Complete	Complete	February 2018	n/a	Yes
7. Enterprise Censuses and Surveys Enabling (ECaSE) – Internet Self Response (ISR) Tool that supports self-response data collection by the Internet for respondents and by call center agents on behalf of respondents.	Complete	Complete	March 2018 (1-month delay)	n/a	Yes
8. 2020 Website Website for the 2018 End-to-End Test, the scope encompasses the Test's internet presence needs.	Complete	Complete	June 2018	January 2019	Yes
9. Master Address File/Topologically Integrated Geographic Encoding and Referencing System	Complete	In progress[b]	December 2016	October 2018	Yes

System name and description	Status of development	Status of testing	Expected/ actual first deployment date[a] (delay since August 2017)	Expected/ actual final deployment date[a] (delay since August 2017)	Has at least a portion of the system's functionality been deployed?
Database that contains, manages, and controls a repository of spatial and non-spatial data used to provide extracts to define census operations, provide maps, and support Web applications.					
10. Census Hiring and Employment Check Administrative system that automates the clearance processing of all personnel at Census Bureau Headquarters, the Bureau of Economic Analysis, the Regional Offices, the National Processing Center, and two Computer Assisted Telephone Interview sites.	Complete	In progress[b]	March 2017	June 2018 (4-month delay)[c]	Yes
11. Census Human Resources Information System Web-based personal information tool providing personnel and payroll information on desktops.	Complete	In progress[b]	March 2017	July 2018	Yes
12. Commerce Business System System that collects and reports labor hours and costs for the activities that the National Processing Center performs.	Complete	In progress[b]	March 2017	July 2018	Yes
13. Decennial Applicant, Personnel and Payroll Systems System that supports personnel and payroll administration for temporary, intermittent Census Bureau employees participating in the 2018 End-to-End test.	Complete	In progress	March 2017	July 2018 (11-month delay)[c]	Yes
14. Decennial Service Center Suite of systems to handle all IT service requests initiated by field staff.	Complete	In progress	March 2017	July 2018	Yes
15. Desktop Services Suite of systems that includes chat.	Complete	In progress	March 2017	July 2018	Yes
16. Recruiting and Assessment Tool that provides capabilities for applicant recruiting and the applicant pre-selection assessment process.	Complete	In progress	March 2017	July 2018 (5-month delay)[c]	Yes
17. Identity Management System System used to ensure that the right individuals have access to the right resources at the right times for the right reasons.	Complete	In progress[b]	March 2017	January 2019	Yes
18. Listing and Mapping Application Single instrument that enables field users to capture and provide accurate listing and mapping updates to the Master Address File/Topologically Integrated Geographic Encoding and Referencing Database.	Complete	In progress	July 2017	April 2018 (1-month delay)	Yes

Table 1. (Continued)

System name and description	Status of development	Status of testing	Expected/actual first deployment date[a] (delay since August 2017)	Expected/actual final deployment date[a] (delay since August 2017)	Has at least a portion of the system's functionality been deployed?
19. Mobile Case Management Tool that provides mobile device-level survey case management and dashboards, and manages data transmissions and other applications on the mobile device.	Complete	In progress	July 2017	April 2018 (1-month delay)	Yes
20. Geospatial Services Tool that provides vintage imagery service, internal current imagery service, public current imagery service, mapping services.	Complete	In progress[b]	July 2017	July 2018	Yes
21. Service Oriented Architecture Enterprise software architecture model used for designing and implementing communication between mutually interacting software applications in a service-oriented architecture.	Complete	In progress	July 2017	July 2018	Yes
22. MOJO Optimizer/Modeling Service to optimize the field workers' routes.	Complete	In progress	August 2017	April 2018 (1-month delay)	Yes
23. Integrated Logistics Management System System to manage logistics and resource planning.	Complete	In progress	August 2017	July 2018	Yes
24. National Processing Center Printing Service that provides printing services for low-volume forms and merges static form and variable data, such as printing a standard form with unique addresses.	Complete	In progress[b]	August 2017	July 2018	Yes
25. Automated Tracking and Control Tool that provides customer, employee, and workflow management by automating business and support activities. It provides outbound call tracking for Geographic Partnership Programs and material tracking and check-in.	Complete	In progress	February 2018	July 2018	Yes
26. Concurrent Analysis and Estimation System System that stores data and uses it to execute statistical models in support of survey flow processing, analysis, and control.	Complete	In progress[d]	March 2018	n/a	Yes
27. Census Questionnaire Assistance Provides call center capability for self-response and assists respondents with responding to and completing census questionnaires.	Complete	In progress	March 2018 (1-month delay)	April 2018 (1-month delay)	Yes
28. Decennial Physical Access System System that maintains the photo and other information relating to providing	Complete	In progress[b]	March 2018	June 2018	Yes

System name and description	Status of development	Status of testing	Expected/actual first deployment date[a] (delay since August 2017)	Expected/actual final deployment date[a] (delay since August 2017)	Has at least a portion of the system's functionality been deployed?
physical access control to facilities. It also is used to generate badges for certain employees, including enumerators, listers, and Census Field Supervisors.					
29. Census Image Retrieval Application Application that provides secure access to census data and digital images of the questionnaires from which the data were captured.	Complete	In progress[b]	March 2018 (1-month delay)	July 2018 (5-month delay[c])	Yes
30. Integrated Computer Assisted Data Entry Tool that captures paper responses from questionnaires.	Complete	In progress	March 2018 (1-month delay)	July 2018	Yes
31. Unified Tracking System Data warehouse that combines data from a variety of Census systems, bringing the data to one place where the users can run or create reports to analyze survey and resource performance.	In progress	In progress	December 2016	October 2018 (7-month delay[c])	Yes
32. ECaSE – Field Operational Control System System that manages field assignments with routing optimizer, reviews and approves field worker's time and expense, and tracks field worker's performance.	In progress	In progress	July 2017	July 2018 (4-month delay[c])	Yes
33. ECaSE Operational Control System System that manages the data collection universe for all enumeration operations, maintains operational workloads, and provides alerts to management.	In progress	In progress	August 2017	July 2018 (4-month delay[c])	Yes
34. Sampling, Matching, Reviewing, and Coding System System that supports quality control for field operations.	In progress	In progress	August 2017	October 2018 (7-month delay[c])	Yes
35. Intelligent Postal Tracking Service Mail tracking system developed by the Census Bureau and the U.S. Postal Service system to trace individual mail pieces during transit.	In progress	In progress	February 2018	July 2018	Yes
36. Control and Response Data System System that provides a sample design and universe determination for the Decennial Census	In progress	In progress	February 2018	October 2018	Yes
37. Census Data Lake Repository for response data that provides data access to reporting and analytics applications.	In progress	In progress	February 2018	January 2019	Yes
38. Decennial Response Processing System System that performs data processing on the raw response data and stores the final processed response data for long term storage.	In progress	In progress	March 2018 (1-month delay)	January 2019	Yes
39. Production Environment for Administrative Records Staging, Integration and Storage	In progress	In progress	March 2018 (1-month delay)	January 2019 (10-month delay[c])	Yes

Table 1. (Continued)

System name and description	Status of development	Status of testing	Expected/ actual first deployment date[a] (delay since August 2017)	Expected/ actual final deployment date[a] (delay since August 2017)	Has at least a portion of the system's functionality been deployed?
Tool that manages Administrative Records and provides services associated with those records.					
40. ECaSE – Enumeration Tool that captures survey responses collected by door-to-door enumeration, records contact attempts, and collects employee availability and time and expenses.	In progress	In progress	March 2018 (1-month delay)	July 2018 (4-month delay)[c]	Yes
41. Centurion Tool that provides an external interface for the upload of group quarters electronic response data.	In progress	In progress	July 2018 (4-month delay)	n/a	No
42. Fraud Detection System System that identifies fraudulent responses either in real-time or post data collection.	In progress	In progress	October 2018 (8-month delay)	n/a	No
43. Center for Enterprise Dissemination Services and Consumer Innovation Tool that will provide search and access to tabulated Census data.	In progress	In progress	January 2019	n/a	No
44. Tabulation Tool that receives post-processed response data and produces tabulated statistical data.	In progress	In progress	January 2019	n/a	No

Key: n/a = not applicable. These systems are only being deployed one time, so the first deployment date also represents the final deployment date.
Source: GAO analysis of Census Bureau data. | GAO-18-543T

[a] The dates listed for March 2018 or earlier should be considered actual dates.

[b] Bureau officials stated that testing for this system is complete; however, the Bureau has not yet provided documentation to support this assertion.

[c] According to officials within the Bureau's 2020 Census Systems Engineering and Integration office, the delay in the final deployment date for this system is due to a change in the timing of the operations it is supporting for the 2018 End-to-End Test.

[d] Although this system has deployed, one of its interfaces was still undergoing testing as of the end of March 2018.

According to the Bureau, a single system may be deployed multiple times throughout the test (with additional or new functionality) if that system is needed for more than one of these operations.

Table 1 describes the status of development and testing, and describes if a portion of functionality has been deployed for each system in the 2018 End-to-End Test. The table also describes key system deployment dates and the delay in these dates since August 2017.

INDEX

#

2020 Census, v, vii, viii, ix, 1, 2, 3, 5, 6, 9, 11, 12, 13, 14, 15, 16, 18, 19, 20, 21, 22, 25, 26, 27, 28, 29, 31, 32, 35, 36, 37, 38, 39, 40, 41, 42, 43, 44, 45, 46, 47, 48, 49, 50, 52, 53, 54, 55, 56, 57, 58, 59, 60, 61, 62, 63, 64, 66, 68, 69, 70, 71, 72, 74, 75, 77, 79, 80, 81, 83, 84, 90, 91, 93, 94, 95, 96, 97, 98, 101, 102, 103, 104, 105, 106, 107, 108, 111, 112, 113, 115, 118, 120, 121, 122, 123, 124, 125, 126, 127, 130, 131, 133, 134, 136, 137, 142

A

access, 8, 38, 66, 67, 68, 70, 71, 126, 139, 141, 142
agencies, 45, 46, 50, 62, 66, 67, 70, 80, 127

B

Bureau of Labor Statistics, 29
businesses, 11, 12, 13, 39, 106

C

Census, vii, viii, ix, 1, 2, 3, 4, 5, 6, 7, 8, 9, 10, 11, 12, 13, 14, 15, 16, 17, 18, 19, 20, 21, 22, 23, 25, 26, 27, 28, 29, 30, 31, 32, 35, 36, 37, 38, 39, 40, 41, 42, 43, 44, 45, 46, 47, 48, 49, 50, 51, 52, 53, 54, 55, 56, 57, 58, 59, 60, 61, 62, 63, 64, 66, 68, 69, 70, 71, 72, 73, 74, 75, 76, 77, 78, 79, 80, 81, 103, 104, 105, 106, 107, 108, 109, 110, 111, 112, 113, 114, 115, 116, 117, 118, 119, 120, 121, 122, 123, 124, 125, 126, 127, 128, 129, 130, 131, 132, 133, 134, 135, 136, 137, 138, 139, 140, 141, 142
census data, vii, viii, 2, 7, 14, 24, 39, 44, 49, 54, 55, 60, 99, 104, 106, 109, 114, 120, 126, 141, 142
challenges, viii, ix, 2, 3, 4, 5, 6, 29, 35, 36, 37, 38, 40, 44, 52, 54, 55, 58, 60, 61, 62, 63, 65, 68, 70, 71, 78, 79, 80, 81, 104, 105, 106, 107, 109, 111, 112, 114, 115, 118, 121, 122, 123, 126, 127, 129, 132, 134, 137
children, 3, 5, 7, 8, 10, 11, 15, 16
citizenship, 44, 45, 57, 109
communication, 25, 69, 140
communities, 3, 4, 8, 11, 13, 21, 30, 48
Congress, iv, 7, 11, 45, 50, 61, 81, 125, 137

Consolidated Appropriations Act, 12, 45, 46, 47, 109, 110
constitution, vii, viii, ix, 35, 39, 44, 88, 103, 106
contingency, 37, 42, 43, 56, 57, 71, 75, 76, 77, 131
coordination, 66, 67, 115, 126
cost, viii, ix, 5, 6, 19, 20, 22, 35, 36, 37, 38, 39, 40, 42, 43, 44, 50, 52, 53, 58, 69, 72, 73, 74, 75, 76, 77, 78, 79, 80, 81, 103, 104, 105, 106, 107, 108, 109, 112, 113, 115, 123, 124, 125, 126, 130, 131, 132, 133, 134, 135, 136, 137
cybersecurity, viii, 36, 38, 40, 41, 52, 57, 61, 62, 63, 65, 66, 67, 68, 71, 80, 81, 131
cybersecurity risks, viii, 36, 41, 52, 61

D

data center, 50, 122, 128
data collection, 24, 44, 54, 55, 60, 109, 111, 113, 138, 141, 142
data processing, 141
decennial census, vii, viii, ix, 35, 36, 39, 40, 43, 44, 50, 51, 69, 71, 78, 79, 88, 93, 95, 101, 103, 104, 106, 107, 109, 111, 123, 128, 132, 135, 141
decentralization, 26
decision makers, 57
deficiencies, 37, 62, 65, 66, 127, 129
demographic change, 39, 106
Department of Commerce, 2, 32, 36, 39, 45, 105, 106
Department of Homeland Security, 38, 41
Department of Justice, 45
detection, 61, 67, 118, 120, 122

E

employees, 49, 61, 69, 70, 121, 139, 141

engineering, 48, 50, 53, 62, 69, 104, 112, 123
environment, 29, 48, 49, 70, 71, 120, 128
estimation process, 38, 50, 74, 79, 106, 132
evidence, 6, 41, 44, 45, 67, 108
executive branch, 50, 62, 66, 127

F

federal assistance, 7
federal law, 127
fiscal year, 12, 19, 20, 23, 42, 43, 45, 47, 75, 108, 110, 113, 123
focus groups, 9, 14, 15, 26, 27
fraud, 40, 61, 107, 118, 120, 122
funding, 5, 11, 12, 17, 20, 29, 47, 54, 77, 111, 114, 128
funds, 23, 24, 42, 43, 77, 125

G

governance, 24, 115, 125, 126
governments, 11, 12, 23, 54, 113
growth, 77, 105, 123, 125
guidance, 16, 55, 56, 62, 79, 127

H

hard-to-count, vii, viii, 2, 3, 5, 6, 8, 9, 13, 14, 15, 16, 24, 26, 27, 28, 29, 30, 31, 32, 48
hard-to-count groups, vii, viii, 2, 5, 6, 8, 13, 15, 16, 24, 26, 29, 30, 31
high school diploma, 8
high-risk list, vii, viii, ix, 36, 40, 50, 51, 52, 104, 107
hiring, 2, 3, 4, 6, 11, 20, 22, 23, 28, 29, 30, 31, 32, 47, 48, 120, 122
House of Representatives, 5, 39, 106

housing, 8, 14, 15, 24, 44, 47, 55, 76, 105, 109, 114, 117, 118, 119, 138

I

individuals, 3, 9, 13, 17, 38, 49, 57, 66, 70, 125, 126, 139
information sharing, 67
information technology, 36, 40, 105, 106
infrastructure, 37, 47, 49, 50, 60, 67, 70, 72, 111, 121, 122, 123, 128
instructional materials, 3, 16
integration, 24, 26, 27, 41, 50, 57, 59, 60, 61, 79, 108, 115, 118, 122, 123, 124, 134
issues, 17, 36, 40, 44, 46, 52, 55, 61, 65, 78, 104, 106, 109, 110, 125, 132, 136
IT systems, viii, 35, 36, 37, 40, 41, 44, 49, 50, 52, 58, 60, 61, 68, 71, 79, 81, 85, 98, 99, 105, 107, 108, 109, 111, 112, 115, 137

L

labor market, 4, 29, 30
language barrier, 11
languages, 3, 5, 17, 18, 48
leadership, 50, 51, 52, 67, 78, 79, 133, 134
life cycle, 6, 19, 20, 127
living arrangements, 5
local community, 11
local government, 13, 134

M

management, 3, 4, 24, 25, 26, 27, 28, 30, 31, 37, 44, 50, 54, 55, 57, 58, 61, 67, 77, 78, 79, 81, 109, 113, 115, 122, 123, 127, 132, 133, 134, 137, 140, 141
materials, 16, 17, 19, 22, 26, 45, 125, 138
measurement, 55, 114, 120, 121, 122

mobile device, 17, 49, 50, 54, 70, 114, 121, 123, 124, 140
mobile phone, 124
modified IT systems, viii, ix, 35, 103

N

natural disaster, 26, 76
nursing home, 15, 118

O

officials, viii, 2, 4, 6, 9, 12, 14, 17, 19, 20, 22, 23, 24, 26, 27, 29, 30, 39, 41, 42, 45, 47, 48, 49, 57, 58, 70, 75, 77, 78, 81, 108, 111, 118, 119, 121, 125, 127, 128, 133, 135, 136, 142
operations, 3, 5, 13, 14, 16, 24, 25, 26, 27, 31, 32, 37, 40, 42, 43, 47, 49, 50, 53, 54, 55, 57, 58, 59, 60, 61, 62, 70, 71, 74, 77, 79, 104, 105, 107, 111, 112, 113, 114, 116, 117, 118, 119, 120, 121, 122, 124, 126, 127, 137, 139, 141, 142
organizational culture, 78, 132
outreach, 3, 11, 13, 14, 15, 19, 20, 21, 23, 26, 30, 48
oversight, 18, 50, 79, 81, 123, 125, 126, 133, 134, 137

P

population, vii, viii, ix, 2, 5, 7, 8, 13, 16, 17, 23, 26, 35, 39, 42, 44, 52, 53, 70, 86, 91, 103, 106, 108, 109, 112
population group, 26
preparation, iv, 12, 23, 66, 78, 132
primary function, 76
private information, 127

Q

questionnaire, 3, 8, 11, 16, 18, 45, 52, 76, 112, 117, 119, 124

R

recommendations, iv, 2, 6, 32, 36, 38, 39, 40, 41, 53, 54, 57, 62, 66, 67, 68, 71, 78, 80, 81, 104, 106, 107, 112, 114, 126, 131, 133, 136
recruiting, 4, 29, 30, 47, 120, 138, 139
reliability, 6, 38, 70, 72, 131
requirements, 24, 57, 61, 62, 70, 76, 127
resources, 17, 22, 24, 26, 44, 45, 46, 51, 109, 110, 127, 139
response, 3, 13, 14, 15, 16, 17, 22, 26, 27, 37, 38, 41, 44, 49, 50, 53, 54, 58, 59, 60, 63, 66, 67, 69, 70, 71, 74, 76, 80, 109, 111, 112, 113, 114, 117, 119, 120, 121, 122, 124, 129, 131, 136, 137, 138, 140, 141, 142
risk, ix, 28, 30, 31, 36, 37, 40, 41, 50, 51, 52, 53, 56, 57, 58, 59, 60, 61, 62, 63, 65, 66, 67, 68, 69, 76, 77, 78, 79, 80, 81, 104, 107, 111, 112, 113, 114, 115, 127, 128, 131, 133, 134, 135, 137
risk management, 37, 40, 57, 62, 63, 66, 67, 68, 78, 127, 133

S

scale system, 49, 120
schedule delays, 61, 79, 123, 126
scope, 5, 24, 31, 41, 55, 107, 114, 120, 122, 138
Secretary of Commerce, 32, 44, 76, 79, 80, 109, 125, 126, 133, 136
security, ix, 36, 37, 38, 41, 61, 62, 63, 64, 65, 66, 67, 70, 71, 72, 79, 80, 81, 104, 105, 106, 108, 115, 118, 119, 120, 121, 126, 127, 128, 129, 136, 137
security practices, 71, 80, 106, 136
security threats, 61, 67
services, iv, 15, 22, 39, 68, 106, 118, 122, 124, 138, 140, 142
sociodemographic groups, vii, 2, 7, 8, 13
specialists, 3, 12, 19, 20, 22, 23, 29, 30, 31, 48
staffing, 12, 17, 21, 31, 48, 61, 122, 134
statutory provisions, 125
successful census, vii, viii, 36, 52, 112

T

technical comments, 2, 27, 32, 41, 108
technical support, 122
technological change, 63, 129
technology, 49, 50, 53, 55, 61, 111, 113, 114, 122, 126, 133
testing, viii, 5, 6, 26, 27, 28, 36, 37, 40, 41, 45, 49, 50, 53, 55, 58, 59, 60, 61, 67, 78, 79, 80, 81, 104, 106, 108, 109, 113, 114, 115, 116, 117, 118, 119, 120, 121, 122, 123, 126, 129, 132, 133, 136, 138, 139, 140, 141, 142, 143
time frame, 23, 38, 42, 43, 56, 57, 60, 65, 71, 75, 113, 118, 119, 126
Title I, 45, 47, 62, 110, 127
Title II, 62, 127
transformation, 40, 107

U

U.S. Department of Commerce, 71
U.S. population, vii, viii, ix, 35, 39, 103, 106
United States, 1, 4, 35, 42, 103, 108, 134

W

wage level, 131
wage rate, 42, 77
workers, 13, 44, 47, 109, 131, 140
workflow, 140
working groups, viii, 2, 6

Related Nova Publications

KEY CONGRESSIONAL REPORTS FOR FEBRUARY 2019. PART II

EDITOR: Mandy Todd

SERIES: Congressional Policies, Practices and Procedures

BOOK DESCRIPTION: This book is a comprehensive compilation of all reports, testimony, correspondence and other publications issued by the Congressional Research Service during the month of February, grouped according to topics. This book is focused on the following topics:
- Business
- Finance

HARDCOVER ISBN: 978-1-53615-737-6
RETAIL PRICE: $195

KEY CONGRESSIONAL REPORTS ON INTERNATIONAL AFFAIRS

EDITOR: Hattie Ross

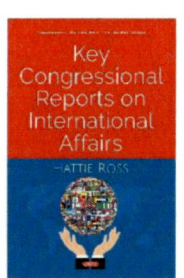

SERIES: Congressional Policies, Practices and Procedures

BOOK DESCRIPTION: This book is a comprehensive compilation of all reports, testimony, correspondence and other publications issued by the Congressional Research Service on U.S. International Relations during the month of February.

HARDCOVER ISBN: 978-1-53615-733-8
RETAIL PRICE: $230

To see a complete list of Nova publications, please visit our website at www.novapublishers.com

Related Nova Publications

Key Congressional Reports for February 2019. Part III

Editor: Mandy Todd

Series: Congressional Policies, Practices and Procedures

Book Description: This book is a comprehensive compilation of all reports, testimony, correspondence and other publications issued by the Congressional Research Service during the month of March, grouped according to topics.

Hardcover ISBN: 978-1-53615-997-4
Retail Price: $230

Key Congressional Reports for March 2019. Part I

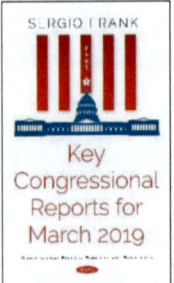

Editor: Sergio Frank

Series: Congressional Policies, Practices and Procedures

Book Description: This book is a comprehensive compilation of all reports, testimony, correspondence and other publications issued by the Congressional Research Service during the month of March, grouped according to topics. This book is focused on the following topics:
· Information Security & Technology
· Military Technology

Hardcover ISBN: 978-1-53615-976-9
Retail Price: $160

To see a complete list of Nova publications, please visit our website at www.novapublishers.com